Studies in International Law of the Sea and Maritime Law
Internationales Seerecht und Seehandelsrecht

Herausgegeben von / edited by

Prof. Dr. Doris König
Prof. Dr. Nele Matz-Lück
Prof. Dr. Alexander Proelß
Prof. Trine-Lise Wilhelmsen

Band / vol. 9

Dr. Pablo Ferrara, LL.M.

Sovereignty Disputes and Offshore Development of Oil and Gas

Nomos

The Deutsche Nationalbibliothek lists this publication in the
Deutsche Nationalbibliografie; detailed bibliographic data
are available on the Internet at http://dnb.d-nb.de

ISBN   978-3-8487-3101-5 (Print)
        978-3-8452-7174-3 (ePDF)

British Library Cataloguing-in-Publication Data
A catalogue record for this book is available from the British Library.

ISBN   978-3-8487-3101-5 (Print)
        978-3-8452-7174-3 (ePDF)

Library of Congress Cataloging-in-Publication Data
Ferrara, Pablo
Sovereignty Disputes and Offshore Development of Oil and Gas
Pablo Ferrara
187 p.
Includes bibliographic references.

ISBN   978-3-8487-3101-5 (Print)
        978-3-8452-7174-3 (ePDF)

1. Edition 2016
© Nomos Verlagsgesellschaft, Baden-Baden, Germany 2016. Printed and bound in Germany.

To Mariana A. Miglino

With all my love,
and gratitude,
for joining me
at adventurous journeys
throughout unknown seas.

Co-editors preface

In a time of reviving conflicts between States about access to and allocation of marine non-living resource deposits, Dr. Ferrara's book, by analyzing and assessing varying designs of bilateral cooperation schemes on the exploitation of offshore oil and gas as well as their subsequent dynamics, addresses a highly topical, practically relevant and scientifically ambitious topic. As far as the methods used are concerned, the study is based on a combined and particularly innovative approach, bringing together the disciplines of international law and international relations. While Dr. Ferrara's findings will not always be approved unanimously, they most certainly have the potential to strongly impact the debate on and understanding of cooperation in the exploitation of disputed resource deposits. This is particularly due to the fact that the study focusses on bilateral cooperation schemes in the Middle East, the Gulf of Mexico and West Africa – schemes that have so far not received the deserved attention in international legal and international relations scholarship. It is thus hoped that Dr. Ferrara's book will be well-received by all those interested in offshore oil and gas cooperation regimes

Trier, April 2016   *Alexander Proelss*

Acknowledgements

> *'I promised myself to go the same way when spring really bloomed and opened up the land.*
> *And this was really the way that my whole road experience began, and the things that were to come are too fantastic not to tell.'*
>
> Jack Kerouac (1957)

Spring was gone and it was already autumn when I arrived at Berkeley in 2006. I was already lucky. I had contacted Professor Harry N. Scheiber to be his research assistant during my stay at the School of Law... And luck stayed with me for a while, as I was able to study immediately with Professor David D. Caron and work with him later on. Without their trust and support, this study would not have been achieved.

It took me ten years to get to the point where my – *basic* – academic craftsmanship was recognised by those who are beyond mastery in the art. I want to thank to the people who made such an outstanding contribution to my formation at Berkeley's School of Law, namely, Professor Cymie Payne, Professor Neil Popovic, Professor Richard Buxbaum and Professor Anthony Rossmann.

I tenderly thank my parents for the thirty-six years gone by – and for those to come.

Being far away and travelling back home, time and again, is not easy. It is only then that people reveal themselves, in different ways. Among them, I am particularly thankful to Alejandra Mpolas, Juan Carlos Risso Patrón, Romina Pardo, Ruben Vela, Ignacio Nantillo, Toni Mendicino, Matías Luchinsky, Hernan Anllo, Jorge Lapeña and Harry Panagopulos.

Then, I would like to take a step backwards chronologically to thank the professors who helped get me to Berkeley in the first place. They are Professor Dr. Lilian del Castillo de Laborde, Professor Dr. Mónica Pinto, Professor Dr. Juan Sola, Professor Dr. Carlos Strasser, Professor Dr. Roberto Russell, and Professor Dr. Eugenio Raúl Zaffaroni.

H.E. Dr. Hugo Caminos deserves special acknowledgement, by whose example, encouragement and friendship I was able to achieve the goal of my academic profession.

Finally, I would like to thank Prof. D. Alexander Proelß, *factotum* of publishing this work in English.

'Driving may be difficult when it is dark outside,
but a science that tries to see the road ahead by using only
the rearview mirror makes little sense,
especially if we are building the road as we go along'

Alexander Wendt

Contents

List of Maps

List of Tables

List of Images

1 Overture: Cooperation in the exploitation of offshore oil & gas

Why study offshore oil and gas cooperation regimes?

The context of political changes after the end of the Cold War, the expansion of the economy by globalisation, along with evolution of the international law of the sea, international law of natural resources and international environmental law, increased concerns about transboundary offshore oil and gas deposits as well as existing and potential offshore conflicts to the fore. Thus, development of offshore cooperation 'regimes' has received increasing attention over the last forty years among scholars as an alternative to traditional methods of dispute resolution in public international law. In this sense, 'institution' and 'regime' are used interchangeably; the concept can be likened to 'regime' as defined by Oran Young: 'human artefacts whose distinguishing feature is the conjunction of *convergent expectations* and recognized patterns of behaviour or practice in a given issue-area of social relations.'[1]

The present work intents, first, to describe and explain variations in the design of three bilateral cooperation regimes between states with offshore common oil and gas deposits. Secondly, this work aims at analysing whether variations in the design of such cooperation institutions lead to synergic variation in the outcomes, dynamics and evolution of that cooperation.

Lastly, it is worth mentioning that the conception and analysis coming out of the following lines were produced between 2007 and 2010. Any modification of treaties or data after its conceit should not be considered relevant.

---

1 *See* Anne-Marie Slaughter, *International law and international relations theory: a prospectus, in* THE IMPACT OF INTERNATIONAL LAW ON INTERNATIONAL COOPERATION at 27, n. 30 (Eyal Benvenisti and Moshe Hirsch eds., Cambridge University Press 2004). *See* James Fearon and Alexander Wendt, *Rationalism v. Constructivism: A Skeptical View, in* HANDBOOK OF INTERNATIONAL RELATIONS (2007) 52, 54.

2 Framework: Markets, cases and stages

The world economy grew steadily for twenty years since 1988, along with energy consumption (Table 1.1), the price of oil (Table 1.2)[2] and the price of gas (Table 1.3). Even though the economy entered a recession period in 2010, world energy consumption projected an increase of sixty percent between 1997 and 2020 (*see* Picture 1.1).[3] This context intensified concern for energy security of supply,[4] geopolitical tension and confrontation across boundaries.

---

2 Oil reached USD 140 per barrel in early July 2008 – a record even when adjusted for inflation.
3 ENERGY INFORMATION ADMINISTRATION, OFFICE OF ENERGY MARKETS AND END USE, INTERNATIONAL STATISTICS DATABASE AND INTERNATIONAL ENERGY ANNUAL 1997, DOE/ EIA-0219(97) (April, 1999). Image from ENERGY INFORMATION ADMINISTRATION, WORLD ENERGY PROJECTION SYSTEM (2000).
4 The literature has defined energy security 'as a condition in which a nation and all, or most, of its citizens and businesses have access to sufficient energy resources at reasonable prices for the foreseeable future free from serious risk of major disruption of service. At different times, in different countries the crucial elements of energy security will vary. Consider the 1996 Report to the Trilateral Commission, Maintaining Energy Security in a Global Context prepared by William Martin, Ryukichi Imai, and Helga Steeg, three energy veterans with worldwide perspectives. They observed: Energy security has three faces. The first involves limiting vulnerability to disruption given rising dependence on imported oil from an unstable Middle East. The second, broader face is, over time, the provision of adequate supply for rising demand at reasonable prices – in effect, the reasonably smooth functioning over time of the international energy system. The third face of energy security is the energy-related environmental challenge. The international energy system needs to operate within the constraints of "sustainable development" – constraints which, however uncertain and long-term, have gained considerable salience in the energy policy debates in our countries'. Barry Barton, Catherine Redgwell, Anita Ronne, and Ronald N. Zillman, *Introduction, in* ENERGY SECURITY. MANAGING RISK IN A DYNAMIC LEGAL AND REGULATORY ENVIRONMENT 3, 5 (Barry Barton *et al.* eds., 2004).

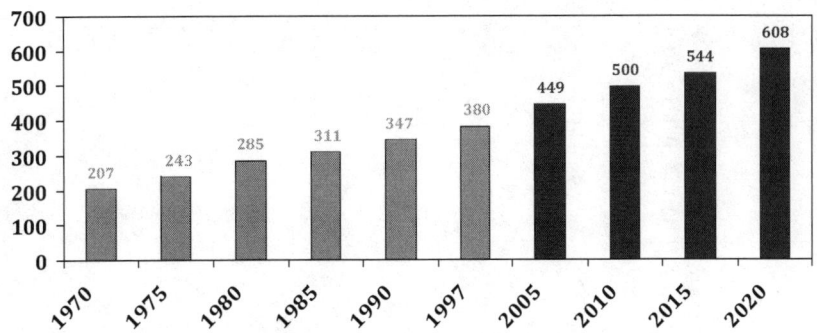

Image 1.1: Projection of Energy Consumption 1997–2020

Million tonnes oil equivalent	1998	1999	2000	2001	2002	2003	2004	2005	2006	2007	2008
World Total	8,888.50	9,021.50	9,262.60	9,323.10	9,502.80	9,810.50	10,258.8	10,555.30	10,820.80	11,104.40	11,294.90

In this Review, primary energy comprises commercially traded fuels only. Excluded, therefore, are fuels such as wood, peat and animal waste which, though important in many countries, are unreliably documented in terms of consumption statistics. Also excluded are wind, geothermal and solar power generation – accounting for less than 0.05%.

Notes: Oil consumption is measured in million tonnes; other fuels in million tonnes of oil equivalent. Growth rates are adjusted for leap years.

Table 1.1: Primary Energy Consumption 1998–2008 (BP 2009)

SPOT CRUDE PRICES				
USD per barrel	**Dubai**	**Brent**	**Nigerian Forcados**	**West Texas Intermediate**
	USD/bbl*	**USD/bbl****	**USD/bbl**	**USD/bbl*****
1988	13.27	14.92	15.00	**15.97**
1989	15.62	18.23	18.30	**19.68**
1990	20.45	23.73	23.85	**24.50**
1991	16.63	20.00	20.11	**21.54**
1992	17.17	19.32	19.61	**20.57**
1993	14.93	16.97	17.41	**18.45**
1994	14.74	15.82	16.25	**17.21**
1995	16.10	17.02	17.26	**18.42**
1996	18.52	20.67	21.16	**22.16**
1997	18.23	19.09	19.33	**20.61**
1998	12.21	12.72	12.62	**14.39**
1999	17.25	17.97	18.00	**19.31**
2000	26.20	28.50	28.42	**30.37**
2001	22.81	24.44	24.23	**25.93**
2002	23.74	25.02	25.04	**26.16**
2003	26.78	28.83	28.66	**31.07**
2004	33.64	38.27	38.13	**41.49**
2005	49.35	54.52	55.69	**56.59**
2006	61.50	65.14	67.07	**66.02**
2007	68.19	72.39	74.48	**72.20**
2008	94.34	97.26	101.43	**100.06**

Source: Platts.

*1972–1985 Arabian Light, 1986–2008 Dubai dated.

**1976–1983 Forties, 1984–2008 Brent dated.

***1976–1983 Posted WTI prices, 1984-2008 Spot WTI (Cushing) prices.

Table 1.2: Spot Crude Oil Prices 1988–2008 (BP 2009)

USD per million Btu	LNG	Natural Gas				Crude Oil
	Japan	European Union	UK	US	Canada	OCDE
	CIF	CIF	Heren NBP Index*	Henry Hub**	Alberta**	CIF
1998	3.34	14.92	-	**-**	-	**2.56**
1989	3.28	18.23	-	**1.70**	-	**3.01**
1990	3.64	23.73	-	**1.64**	1.05	**3.82**
1991	3.99	20.00	-	**1.49**	0.89	**3.33**
1992	3.62	19.32	-	**1.77**	0.98	**3.19**
1993	3.52	16.97	-	**2.12**	1.69	**2.82**
1994	3.18	15.82	-	**1.92**	1.45	**2.70**
1995	3.46	17.02	-	**1.69**	0.89	**2.96**
1996	3.66	20.67	1.87	**2.76**	1.12	**3.54**
1997	3.91	19.09	1.96	**2.53**	1.36	**3.29**
1998	3.05	12.72	1.86	**2.08**	1.42	**2.16**
1999	3.14	17.97	1.58	**2.27**	2.00	**2.98**
2000	4.72	28.50	2.71	**4.23**	3.75	**4.83**
2001	4.64	24.44	3.17	**4.07**	3.61	**4.08**
2002	4.27	25.02	2.37	**3.33**	2.57	**4.17**
2003	4.77	28.83	3.33	**5.63**	4.83	**4.89**
2004	5.18	38.27	4.46	**5.85**	5.03	**6.27**
2005	6.05	54.52	7.38	**8.79**	7.25	**8.74**
2006	7.14	65.14	7.87	**6.76**	5.83	**10.66**
2007	7.73	72.39	6.01	**6.95**	6.17	**11.95**
2008	12.55	97.26	10.79	**8.85**	7.99	**16.76**

*Price is for NBP Day-Ahead Index. Source: ICIS Heren Energy Ltd.

**Source: Natural Gas Week.

Note: Btu = British thermal units; cif = cost+insurance+freight (average prices).

Table 1.3: Gas Prices 1988–2008 (BP 2009)

Legal Contexts: Boundaries Delimitation and Natural Resources Law

In order to begin with this study, it is essential to understand that oil exploration is a high-tech sophisticated and costly business. Consequently, very few companies are willing to risk adventures in disputed areas.

Land-based hydrocarbons in border areas seldom pump up international boundary quarrels as oil companies uphold the informal principle not to acquire concessions in politically-sensitive areas. There are few major onshore oil or gas fields of sufficient interest to industry lying across international boundaries (*see* Map 1.1).[5] In contrast, offshore deposits have generated a particular relevance as many of the most substantial world's reserves of hydrocarbons (proven and potential) lie beneath the ocean (*see* Map 1.2).[6] To address this, the hydrocarbon industry has tried to improve its offshore technology in the last fifty years, increasing exploration and development. Whilst the original focus was on near-shore and shallow-water prospects, it expanded towards higher sea areas where environmental conditions are severe and vulnerable. Altogether, it is possible to assert that the offshore industry is at an early stage of development, but companies are making substantial investments in technology development and exploration/exploitation projects.[7] By 2010, there were more than 6,000 offshore oil and gas installations operating worldwide: 4,000 in the Gulf of Mexico, 950 in Asia, 700 in the Middle East and 400 in Europe.[8]

---

5 BOUNDARIES AND ENERGY: PROBLEMS AND PROSPECTS 3, 6 (Gerald Blake *et al.* eds., 1998).
6 *Id.*, at 9.
7 PETROBRAS ANUNCIÓ INVERSIONES POR U\$S 44.000 MILLONES [Petrobras announced investment of USD 44,000 million], INFOLEG.COM, March 9, 2010, http://www.infobae.com/finanzas/504703-0-0-Petrobras-anunci%F3-inversio nes-us44.000-millones.
8 UNEP Offshore Oil and Gas Environment Forum, http://www.natural-resources.org /offshore/background/bgnote.htm.

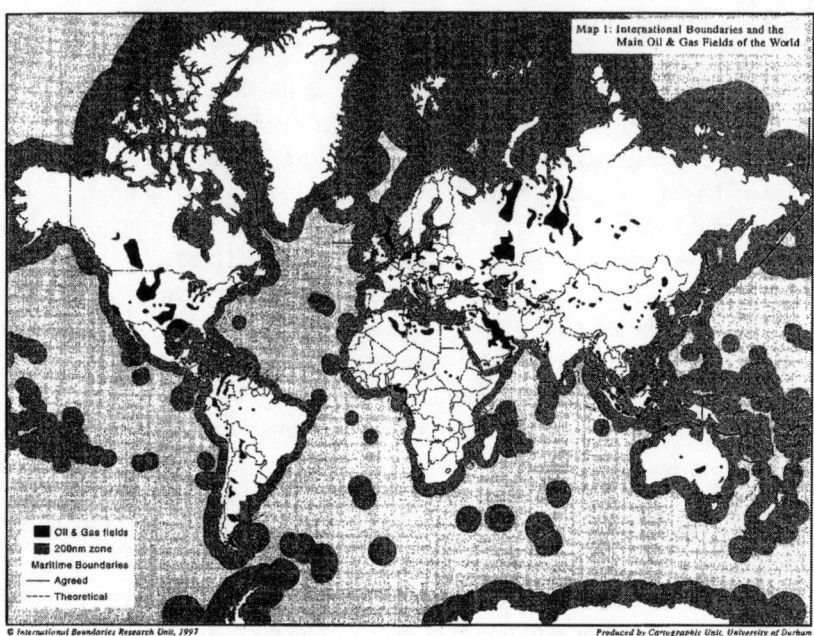

Map 1.1: International Boundaries and the Main Oil and Gas Fields of the World

Now, offshore oil and gas fields can be considered a common resource sometimes. A common resource is defined 'as any natural resource that is used (or is capable of being used) by at least two states'.[9] Shared offshore transboundary petroleum or hydrocarbon deposits pose different legal issues for the sharing states in different instances: i) when the deposit straddles continental shelf boundaries which have been agreed upon and are undisputed; ii) when the deposit lies in a disputed continental shelf area in which two or more states have overlapping claims. Statistics show that the percentage of all maritime borders agreed in 1996 was 35.9 percent,[10] with the remaining extent being considered areas for potential conflict.

---

9 David M. Ong, *Joint Development of Common Offshore Oil and Gas Deposits: "Mere" State Practice or Customary International Law?*, 93 Am. J. INT'L L.771, n. 7.
10 *Supra* note 5, at 11.

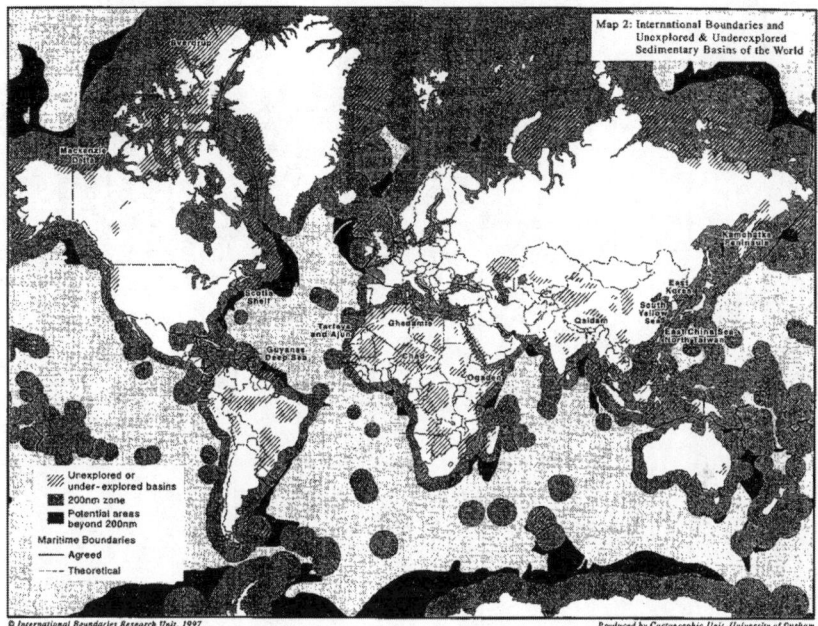

Map 1.2: International Boundaries and Unexplored and Underexplored Basins of the World

Following the different contexts exposed, it is possible to assert two conclusions. First, the development of some offshore areas could satisfy the exponential demand for new sources of oil and gas. Second, cooperation regimes can fulfil states' interests in development and simultaneously conclude or preclude disputes between them.

This study only considers operative regimes with factual consequences. How these criteria are determined will be explained at the end of this chapter. First, previous studies on the topic in question will be described.

Boundaries Delimitation: Cases

| BLURRING |
| JOINT DEVELOPMENT ZONES |
| BOUNDARIES AND ENERGY |
| PEACEFUL MANAGEMENT |
| PETROLEUM INDUSTRY AND GOVERNMENTS |
| LIBRO ESPECIFICO |
| PAPER |

		Year	Parts	Continent	Land/Sea	Boundary Delimitation	Natural Resource	Agreement (Hard/Soft) (Op./Prog.) (System or Form)
1	UNI	1958	Saudi Arabia - Bahrein	Asia	Sea	Yes	HC	Hard/Operative/Joint Development
2	GC	1978	Australia – Papua New Guinea	Oceania	Sea	Yes	HC/Living	Hard/Mixed/HC Moratorium
3	GC	1980/1981	Iceland – Norway (Jan Mayen Island)	Europe	Sea	Yes	HC	Hard/Operative/Joint Zone (Revenues)
4	GC	1999	Denmark – UK	Europe	Sea	Yes	Living	Hard/Operative/Joint Fishing Zone ("Special Area")
5	GC	2000	China – Vietnam (Gulf of Tonkin)	Asia	Sea	Yes	Living	Hard/Operative/Joint Common Fishery Zone

	Year	Parts	Continent	Land/Sea	Boundary Delimitation	Natural Resource	Agreement (Hard/Soft) (Op./Prog.) (System or Form)
6 GC	1982	Cambodia – Vietnam (Gulf of Thailand)	Asia	Sea	No	All (Particularly Fisheries)	Hard/Programmatic/Joint Area ("Historic Waters")
7 UNI	1974	Japan – Korea (East Sea to Korea)	Asia	Sea	No	HC	Hard/Operative/Joint Development Zone
8 JP	1974	Saudi Arabia – Sudan (Red Sea)	Africa – Asia	Sea	No	All (Particularly heavy metals and hot brines)	Hard/Operative/Joint Development Zone
9 UNI	1979	Malaysia – Thailand (Continental Shelf)	Asia	Sea	No	Non living (HC in particular)	Soft (MoU)/Programmatic/ Joint Development Area
10 UNI	1992	Malaysia – Vietnam (Southeast of Gulf of Thailand)	Asia	Sea	No	HC	Hard/Programmatic/Joint Development ("Defined Area")
11 GC	1993	Colombia – Jamaica	America	Sea	Yes	All	Hard/Programmatic/Joint Regime Area
12 JO UNI	1990/1995	Argentina – UK (Malvinas/ Falkland)	America	Sea	No	All	Soft (Joint Statement/Joint Declaration)/Programmatic/ Coordinated Activities
13 UNI	2001	Nigeria – São Tomé and Príncipe	Africa	Sea	No	All (Particularly HC)	Hard/Operative/Joint Development
14 GC UNI	2002/2006	Australia – East Timor (Timor Sea)	Oceania	Sea	No	HC	Hard/Programmatic/Mixed

		Year	Parts	Continent	Land/ Sea	Boundary Delimita- tion	Natural Resource	Agreement (Hard/Soft) (Op./Prog.) (System or Form)
15	GC	1997	China – Japan (East China Sea)	Asia	Sea	No	Fisheries	Hard/ Programmatic/Joint Fishing Zone
16	GC	2001	China – Republic of Korea (Yellow Sea)	Asia	Sea	No	Fisheries	Hard/ Programmatic/Joint Fishing Zone
17	GC	2000	Japan – Republic of Korea (East China Sea and Sea of Japan)	Asia	Sea	No	Fisheries	Hard/ Programmatic/Joint Fishing Zone
18	?	2001	Cambodia – Thailand (Conti- nental Shelf in the Gulf of Thailand)	Asia	Sea	No	Non living	Soft (MoU)/Programmatic/ None
19	UNI	2008	China – Japan (East China Sea)	Asia	Sea	No	All	Soft ("Principled Consensus)/ Mixed/Joint Development
20	-	1975	Colombia – Ecuador	America	Sea	Yes	All	Hard/Programmatic/Principle ("Duty to Cooperate")
21	-	1976	Colombia – Panama	America	Sea	Yes	All	Hard/Programmatic/Principle ("Duty to Cooperate")
22	-	1977	Colombia - Costa Rica	America	Sea	Yes	Migrating Species and Pol- lution Prevention	Hard/Programmatic/Principle ("Duty to Cooperate")
23	-	1977	Cuba – Haiti	America	Sea	Yes	All	Hard/Programmatic/Principle ("Duty to Cooperate")

	Year	Parts	Continent	Land/Sea	Boundary Delimitation	Natural Resource	Agreement (Hard/Soft) (Op./Prog.) (System or Form)
24	-	1978	Colombia - Haiti	America	Sea	Yes	Migrating Species and Pollution Prevention	Hard/Programmatic/Principle ("Duty to Cooperate")
25	-	1978	Netherlands (representing territories of Netherland Antilles: Aruba, Bonaire, Curaçao, Saba. St. Eustatius and St. Maarten) – Venezuela	America	Sea	Yes	HC and Pollution Prevention	Hard/Programmatic/Principle ("Duty to Cooperate")
26	-	1979	Dominican Republic – Venezuela	America	Sea	Yes	Pollution Prevention and Information Exchange	Hard/Programmatic/Principle ("Duty to Cooperate")
27	?	1980	Costa Rica – Panama	America	Sea	Yes	All	Hard/Programmatic/Principle ("Duty to Cooperate")
28	-	1984	Colombia – Costa Rica	America	Sea	Yes	All	Hard/Programmatic/Principle ("Duty to Cooperate")
29	-	1986	Colombia - Honduras	America	Sea	Yes	HC	Hard/Programmatic/Principle ("Duty to Cooperate")
30	-	1990	Trinidad and Tobago – Venezuela	America	Sea	Yes	HC and Pollution Prevention	Hard/Programmatic/Principle ("Duty to Cooperate")

		Year	Parts	Continent	Land/ Sea	Boundary Delimita- tion	Natural Resource	Agreement (Hard/Soft) (Op./Prog.) (System or Form)
31	-	1994	Cuba – Jamaica	America	Sea	Yes	Living Resources and Scientific Research	Hard/Programmatic/Principle ("Duty to Cooperate")
32	-	2001	Honduras – UK (representing the territories of the Cayman Islands)	America	Sea	Yes	Living Resources (Fish-eries)	Hard/Programmatic/Principle ("Duty to Cooperate")
33	UNI	1978	Colombia – Dominican Republic	America	Sea	Yes	Fishing, Scien-tific Research, and Pollution Prevention	Hard/Operative/Zone of Joint Scientific Research and Fish-ing Exploitation
34	UNI	1984	Argentina – Chile ("Sea of the Southern Zone")	America	Sea	Yes	All and Environ-ment Protection	Hard/Operative/Zone Joint Bi-lateral Commission
35	JO	2000	Mexico – US (Gulf of Mex-ico/ "Western Gap")	America	Sea	Yes	HC	Hard/Operative/Joint Zone
36	GC	2003	Barbados – Guyana	America	Sea	Yes	All	Hard/Operative/Co-operation Zone
37	JP	1969	UAE (Abu Dhabi) – Qatar (Al-Bunduq petroleum field at the Persian Gulf)	Asia	Sea	Yes	HC	Hard/Programmatic/Equal sharing and joint sovereignty over HC

		Year	Parts	Continent	Land/Sea	Boundary Delimitation	Natural Resource	Agreement (Hard/Soft) (Op./Prog.) (System or Form)
38	JO	1974	France – Spain	Europe	Sea	?	All	Hard/Programmatic/Equal Sharing of natural resources in a Special Zone
39	JP	1975	Rwanda – Zaire (Lac Kivu)	Africa	Land	?	Gas (Methane)	Hard/Programmatic/Joint Property and Joint Development
40	UNI	1988	Yemen Arab Republic – Democratic Republic of Yemen (now unified as Republic of Yemen)	Asia	Land	Yes	Oil	Hard/Operative/Joint Development Area
41	UNI	1976	Norway – UK (Frigg Reservoir)	Europe	Sea	No	HC	Hard/?/?
42	?	1965	Kuwait – Saudi Arabia	Asia	Land	Yes	HC	?
43	UNI	?	UK – Netherlands (Markham Drill)	Europe	Sea	No	HC	Hard/Operative/Joint Exploitation
44	?	1970/1978	Mexico – US	America	Land/Sea	Yes	?	?
45	GC UNI	1989	Australia – Indonesia	Asia – Oceania	Sea	No	HC	Hard/Operative/Mixed
46	GC	1978	Norway – Russia (Barents Sea)	Europe	Sea	No	Fisheries	Hard/Operative/Joint Fishing Regime ("Grey Zone")

		Year	Parts	Continent	Land/Sea	Boundary Delimitation	Natural Resource	Agreement (Hard/Soft) (Op./Prog.) (System or Form)
47	JO	1962	Germany – Netherlands	Europe	Sea (Estuary of River Ems)	No	Gas	Hard/Operative/Joint Operation
48	GC	1960	Czechoslovakia – Austria (Vysoka –Zwendorf Frontier Region)	Europe	Land	No	Gas	Hard/Operative/Joint Operation
49	UNI	1979	Malaysia – Thailand (Gulf of Thailand)	Asia	Sea	No	HC	Hard/Operative/Joint Development
50	UNI	1993	Guinea Bisseau – Senegal	Africa	?	?	?	Hard/Operative/Joint Development
51	GC	1961/1990	Guatemala – Mexico	America	Sea	Yes	All	Hard/Operative/Joint Development
51	GC JO	1973/1988	Argentina – Uruguay	America	Land/Sea	Yes	All	Hard/Operative/Mixed
52	GC JO	1993	Colombia – Jamaica	America	Sea	Yes	All	Hard/Operative/Mixed
64	Trad.	1993	France (Guadaloupe & Martinique) – UK (Montserrat)	America	Sea	Yes	-	Hard
65	Trad.	1996	Dominican Republic – UK (Turcs and Caicos)	America	Sea	Yes	-	Hard

Natural resources and international oil and gas cooperation: Doctrine and Codification

The study of international oil and gas cooperation regimes took off framed intro the study of law of shared natural resources (aquifers, in particular) as well as the law of boundary delimitation.[11] In this regard, it is possible to divide the literature in three stages and to sum up the codification work in the work of the United Nations International Law Commission.

Doctrine

As posed, the first stage to study international oil and gas cooperation regimes started in the late 60s, following the 1958 Convention on the Continental Shelf[12] and after the first treaties on the subject had already been signed.[13] The second stage emerged after the 1982 United Nations Con-

---

11 Before the Second World War, literature on oil and gas exploitation focused mostly on intra-state regimes applicable only to private subjects. See Ely, *The Conservation of Oil*, 51 Harv. L. Rev. 1209 (1937–1938); *see also* H. WILLIAMS, R. MAXWELL & C. MEYERS, CASES AND MATERIALS ON THE LAW OF OIL AND GAS 1–12 (3rd ed. 1974); cited in Rainer Lagoni, *Oil and Gas Deposits Across National Frontiers*, 73 Am. J. Int'l Law 215, 217 n. 8.

12 THE GENEVA CONVENTION ON THE CONTINENTAL SHELF, U.N. Doc. A/Conf. 13/L.55.

13 A chronological, but not exhaustive, list of these agreements includes the following: AGREEMENT CONCERNING THE DELIMITATION OF THE CONTINENTAL SHELF IN THE PERSIAL GULF, Saudi Arabia–Bahrain, Feb. 22, 1958, NATIONAL LEGISLATION AND TREATIES RELATING TO THE LAW OF THE SEA 409, U.N. Doc. ST/LEG/SER.B/16 (hereinafter LEGISLATION AND TREATIES); AGREEMENT CONVERNING THE WORKING OF COMMON DEPOSITS OF NATURAL GAS AND PETROLEUM, Czech Republic–Austria, Jan. 23, 1960, 495 UNTS 134; SUPPEMENTARY AGREEMENT CONCERNING ARRANGEMENTS FOR COOPERATION IN THE EMS ESTUARY, Netherlands–FRG, May. 14, 1962 509 UNTS 140; AGREEMENT RELATING TO THE PARTITION OF THE NEUTRAL ZONE, Kuwait–Saudi Arabia, Jul. 7, 1965, 4 ILM 1134 (1965); AGREEMENT ON THE SETTLEMENT OF MARITIME BOUNDARY LINES AND SOVEREIGN RIGHTS OVER ISLANDS, Qatar–Abu Dhabi, Mar. 20, 1969, LEGISLATION AND TREATIES 403; MEMORANDUM OF UNDERSTANDING, Iran–Sharjah, Nov. 18, 1971, reprinted in ALI A. EL-HAKIM, THE MIDDLE EASTERN STATES AND THE LAW OF THE SEA 208 (1979); CONVENTION ON THE DELIMITATION OF THE CONTINENTAL SHELF IN THE BAY OF BISCAY, France–Spain, Jan. 29,

vention on the Law of the Sea,[14] slowly evolving through the 1980s and reaching its peak in mid-90s. Eventually, in the early twenty-first century, the second stage split into two different lines of doctrine as a third stage, further involved with the general law of natural resources than with oil and gas in particular.

First Stage

The first stage of literature on oil and gas offshore cooperation regimes was signalled by the works of William T. Onorato and Rainer Lagoni,[15]

1974, LEGISLATION AND TREATIES 445; AGREEMENT CONCERNING JOINT DEVELOPMENT OF THE SOUTHERN PART OF THE CONTINEN-TAL SHELF ADJACENT TO THE TWO COUNTRIES, Japan–South Korea, Feb. 5, 1974, in 4 NEW DIRECTIONS IN THE LAW OF THE SEA 117 (R. R. Churchill & Myron H. Nordquist eds., 1975); AGREEMENT RELATING TO THE EXPLOITATION OF THE SEA-BED AND SUBSOIL OF THE RED SEA IN THE COMMON ZONE, Sudan–Saudi Arabia, May 16, 1974, LEGISLA-TION AND TREATIES, at 452, U.N. Doc. ST/LEG/SER.B/18, U.N. Sales No. E/F.76.V.2 (1976); AGREEMENT RELATING TO THE EXPLOITATION OF THE FRIGG RESERVOIR AND THE TRANSMISSION OF GAS THERE-FROM TO THE UNITED KINGDOM, United Kingdom–Norway, May 10, 1976, 1098 UNTS.

14 UNITED NATIONS CONVENTION FOR THE LAW OF THE SEAS, opened for signature Dec. 10, 1982 1833 UNTS 397, reprinted in UNITED NATIONS, OFFI-CIAL TEXT OF THE UNITED NATIONS CONVENTION ON THE LAW OF THE SEA WITH ANEXES AND INDEX, UN Sales No. E.83.V.5 (1983) (entered into force Nov. 16, 1994); hereinafter UNCLOS.

15 *See generally* William T. Onorato, *Apportionment of an International Common Petroleum Deposit*, 17 INT'L & COMP. L.Q. 85 (1968); William T. Onorato, *Apportionment of an International Common Petroleum Deposit*, 26 INT'L & COMP. L.Q. 324 (1977); Rainer Lagoni, *supra* n. 15; J. E. Horigan, Uni-tization of Petroleum Reservoirs Extending Across Sub-sea Boundary Lines of Bordering States in the North Sea, Natural Resources Lawyer, Vol. 7, No. 1, Win-ter 1974, cited in William T. Onorato, *Apportionment,* (1977), at 324, 332 n.22; Jacobs, *Unit Operation of Oil and Gas Fields*, 57 YALE L.J. 1207 (1947–1948); Talaat El Ghoneimy, The Legal Status of the Saudi-Kuwait Neutral Zone, 15 INT'L & COMP. L.Q. 690 (1966); Hosni, The Partition of the Neutral Zone, 60 AJIL 735 (1966), Al-Baharna, A Note on the Kuwait-Saudi Arabia Neutral Zone Agreement of July 7, 1965, Relating to the Partition of the Zone, 17 INT'L & COMP. L.Q. 730 (1968); Utton, *Institutional Arrangements for Devel-oping North Sea Oil and Gas*, 9 Virginia J. INT'L L. 66 (1968–1969).

who attempted to uncover which rules applied to the particular case of interstate oil and gas deposits.

From the perspective of social inquiry, the first stage literature used the international law research methodology common at that time (*hereinafter*,

---

16 'Description is far from mechanical or unproblematic since it involves selection from the infinite number of facts that could be recorded. There are several funda-mental aspects of scientific description. One is that it involves inference: part of the descriptive task is to infer information about unobserved facts from the facts we have observed. Another aspect involves distinguishing between that which is systematic about the observed facts and that which is nonsystematic'. KING, KEOHANE AND VERBA, DESIGNING SOCIAL ENQUIRY. SCIENTIFIC INFERENCE IN QUALITATIVE RESEARCH, (1994) 34, 34.

17 For an explanation of institutionalism, *see* Beth A. Simmons and Lisa L. Martin, *International Organizations and Institutions, in* INTERNATIONAL LAW AND INTERNATIONAL RELATIONS, 195. Some authors would also include the tra-ditional legalist methodology in a Realist frame: 'Realists and traditional interna-tional lawyers overlap on all three core assumptions: concerning actors, prefer-ences and the constraints imposed by the international system. They do ultimately diverge, with international lawyers seeking to blunt or alter the implications of a pure Realist analysis, but less than either camp might suspect.

The clearest overlap concerns the relevant criteria for identifying participants in the international system. Both Realists and traditional international lawyers agree that the primary actors are states, and define states as monolithic units identifiable only by the functional characteristics that constitute them as states. Neither would take account of domestic political ideology or structure, or of the multiplicity of sub-state actors that determine state policy at the domestic level. Both would assume that rules governing state behavior apply to all states qua states, without regard to their internal identity. The first-order international legal principles of sovereign equality and exclusive domestic jurisdiction are safeguards of the iden-tity and opacity of the sovereign sphere. International legal rules governing recog-nition and state succession similarly ensure a complete divorce between govern-ments and states.

To post-Westphalian international lawyers, then, states are both the source and the subject of rules governing international relations. What motivates and constrains these states in their relations with one another? '[M]ost international lawyers assume that states have at least some common ends and that they can arrange to achieve them by means other than power. Nevertheless, many aspects of tradi-tional international law tacitly acknowledge the extent to which international rela-tions are power relations'. Anne-Marie Slaughter, *International Law and Interna-tional Relations Theory: a Prospectus, in* THE IMPACT OF INTERNATIONAL LAW ON INTERNATIONAL COOPERATION (1998), 16, 23 and 24. *Also see* Stephen D. Krasner, *Structural Causes and Regime Consequences: Regimes as Intervening Variables, in* INTERNATIONAL LAW AND INTERNATIONAL RELATIONS, 3.

'Grotian' methodology), namely a methodology based on strong descriptive inference,[16] contractually instutionalist[17] and analytically Rationalist.[18]

William T. Onorato expressed that his first work was limited to an academic framework, not only by the few 'international common petroleum deposits', as he defined them,[19] but also by the fact that the existing ones where onshore.[20] Nine years later, he said that the framework had changed as the International Court of Justice had already decided in the North Sea Continental Shelf Cases, and there were two offshore deposits in operation and two under exploration in the area.[21] As already noted, his research was purely descriptive.[22] He focused and provided conclusions on four

---

18 'As used in IR [International Relations] contexts, "rationalism" seems to refer variously to formal and informal applications of rational choice theory to IR questions, to any work drawing on the tradition of microeconomic theory from Alfred Marshall to recent developments in evolutionary game theory, or most broadly to any "positivist" exercise in explaining foreign policy by reference to goal-seeking behavior'. James Fearon and Alexander Wendt, *Rationalism v. Constructivism: A Skeptical View, in* HANDBOOK OF INTERNATIONAL RELATIONS (2007) 52, 54.

19 'An international common petroleum deposit is a single petroleum structure or field which underlies in part the territory of two or more States. Such a deposit may be situated on land or offshore. In either location it is still subject to conflicting national rights of exploitation. As such, it is a potential source of international dispute as to the nature and extent of rights as may be asserted by the several sovereigns under whose territory the pool lies. In such a dispute the main problem is to ascertain the content of the law defining the rights of those States interested in the common reserve'. William T. Onorato, *Apportionment of an International Common Petroleum Deposit*, 17 INT'L & COMP. L.Q. (1968) 85, 85.

20 'These two fields are the Zwernsdorf-Vysoka Natural Gas Field shared by Austria and Czechoslovakia, and portions of the Groningen Natural Gas Field which overlaps the international boundary between the Netherlands and the Federal Republic of Germany in the Ems River estuary'. William T. Onorato, *Apportionment of an International Common Petroleum Deposit*, 17 INT'L & COMP. L.Q. (1968) 85, 94 n. 22.

21 'The Frigg Gas Field and the Statfjord Oilfield, both of which were first discovered on the Norwegian side of the common offshore boundary line with the UK. In addition, the Murchison Oilfield was discovered by Conoco in UK Block 211 /19 and Pan Ocean's potentially commercial Brae Oilfield was discovered in UK Block 16/7, but both were still to be studied'. William T. Onorato, *Apportionment of an International Common Petroleum Deposit*, 26 INT'L & COMP. L.Q. (1977) 324, 324 n. 22. For the decision of the International Court of Justice in the North Sea Continental Shelf Cases, *see* ICJ Reports 1969.

22 *Supra*, n. 20.

topics: (1) on the definition and study of the evolution of the law applicable to the apportionment of border-straddling offshore oil and gas deposits, he concluded that the subjects interested in an oil or gas deposit should cooperate;[23] (2) in analysing where there was an emerging law on ownership regarding deposits, he understood that the issue could only be concluded by formal agreement between the parties;[24] (3) regarding the principles that should apply to the exploitation of border-straddling deposits, he considered that: (a) exploitation should not be done under objection and consequently 'unconsented, unilateral exploitation of an international common petroleum deposit by an interested State [was] unlawful under existing international law'[25] and 'such unlawful action, if taken, would be enjoinable and/or answerable in damages',[26] (b) the method of exploitation should be agreed between the parties involved, (c) there was an obligation to negotiate among the interested parties and (d) there was no international law prescribing express methods for apportionment of common deposits, even if there were 'rules and institutions of international law and private law relevant either directly, by analogy or by synthesis to the question under consideration';[27] and finally (4) on the

---

23 In both works, Onorato discarded the possibility that the creation of a boundary under the 'special circumstance' prescribed by Article 6 of the 1958 Convention on the Continental Shelf could be applicable, as this seemed rather to be directed at national ownership and authority (police power). Originally, he had concluded that there was no international customary rule applicable; however, various reasons ultimately led him to infer that this was be the correct approach. His conclusions were based on the need to 'keep the unity of the deposit' in the terms of 'the general principles of law recognized by civilized nations' which exploited oil and gas, on the basis of existing non-applied and applied bilateral treaties, on decisions of different arbitrations, on principles used for other shared natural resources and on the particular geological characteristics of oil and gas.

24 William T. Onorato, *Apportionment of an International Common Petroleum Deposit*, 17 INT'L & COMP. L.Q. (1968) 85, 99.

25 William T. Onorato, *Apportionment of an International Common Petroleum Deposit*, 26 INT'L & COMP. L.Q. (1977) 324, 328.

26 *Id.*, at 330.

27 Originally (1968) his reasoning in this case was based on 'a provision in the Final Protocol to the Supplementary Agreement to the Ems-Dollard Treaty which provides [in the] Art. 4 that "If one of the Contracting Parties has issued no extraction concession for a part of the frontier area...within a reasonable period of time, the holder of an extraction con- cession granted by the other Contracting Party for that part of the frontier area shall have the right...to dispose of the entire deposit of petroleum or natural gas present in that part of the frontier area before the com-

definition of the legal applicability of the previously mentioned substantive principles and rules on the international plane, he dedicated special attention to this in his second work, as it concerned him in particular. As Assistant to the Vice-President (Legal) of the Standard Oil Company of California, he was aware of the possible methods of apportionment that

---

mencement of extraction operations.'" This led him to conclude that '[a] reasonable interpretation of this provision is that the basic rule of equal sharing of produce in kind has been waived here to allow the operator with initiative to gain the benefit of an advantageous contract for disposition of all produce without restraint and uncertainty from the other side because of delay. There is no reason, however, to suppose that the rule of equal sharing of profit has also been waived if equal sharing of costs is subsequently offered. In effect, initiative here creates a greater right of at least economic freedom'. William T. Onorato, *Apportionment of an International Common Petroleum Deposit*, 17 INT'L & COMP. L.Q. (1968) 85, 100, n. 47. Later on, in his second work (1977) he adopted some definitive principles and rules of law to be applied in the abovementioned situation: (i) a state or states interested in an international common petroleum deposit may not unilaterally exploit such a deposit over the reasonable objection of another such state or states; (ii) the method of exploitation along with the underlying legal basis for apportionment of such a deposit must be the object of an agreement between such interested states; (iii) States so interested in an international common petroleum deposit are under an obligation to enter into negotiations with a view toward arriving at such an agreement on its apportionment between them; and (iv) as regards the substance of such negotiations and the principles and rules of law applicable to any such agreement on apportionment of a given reserve, while it is recognised that there is no developed, crystallised rule of international law which prescribes express methods for apportionment of an international common petroleum deposit between interested States, there are, nonetheless, rules and institutions of international law and private law relevant either directly, by analogy or by synthesis to the question under consideration, which rules and institutions include: provisions of the municipal law of oil-producing nations relating to apportionment of common petroleum deposits with particular reference to the established principle of 'unitisation' or 'unit production'; provisions of international law relating to apportionment of international common natural resources having physical properties analogous to those of petroleum; and principles and rules on apportionment of international common petroleum deposits emerging from existing State practice which might be considered 'general practices accepted as law' on the question or which reflect the *opinio juris* in the matter of apportionment, and which thus provide an indication of the policy and principles upon which agreements for apportionment of particular international common petroleum reserves can be voluntarily formulated between interested States, or, alternatively: and which encompass the relevant principles and rules of law in the light of which the methods for eventually effecting the apportionment of a common reserve between interested States will have to be chosen.

could be used for common reserves. Among 'lesser' forms of cooperation, he concluded that '[u]nitisation [was] clearly the best and, accordingly, the prime objective to aim for'.[28] He had already asserted:

> [O]ne must assume that *fair and equitable principles must necessarily be applied.* As one recent author noted,[29] in terms of the present state of knowledge in the petroleum industry, this means that decisions to be taken on an *equitable apportionment* of a common reserve *should be based on* sound petroleum engineering principles and, in addition, *some form of unitised operation to assure the highest degree of effective and efficient co-operation.*[30]

Finally, Onorato supported his conclusion by referring to existing interstate agreements.[31] He was most interested to underline the precise international legal corpus applicable to existing common oil deposits. From state practice, he understood that joint-mutual development of common reserves was the best form of cooperation among states, and that it would be even better to unify-codify technical procedures.[32] To him, *de lege* implied *de facto*.

Rainer Lagoni wrote a more elaborated work a few years after Onorato. He stepped outside pure Grotian doctrine and addressed, albeit narrowly, some of the factual consequences that cooperation regimes could have. He also recognised that private actors could affect decisions taken in the international arena.[33]

---

28 'This form of unitised operation, or "unitisation," may be simply defined as the process whereby separate interest owners in a common reserve pool such interests to form a single unit under the sole operation of a single operator who conducts unit operations for all so that maximum efficient recovery is accomplished and production and/or revenues therefrom may be shared out in accordance with the agreed basis established in the unit plan. The location of platforms, production facilities and the spacing, number and production rates of wells are regulated by the unit operator in an orderly and efficient manner to assure such maximum efficient recovery of the reserve without unnecessary waste or the drilling of uneconomic wells from duplicate platforms or other production facilities'. William T. Onorato, *Apportionment of an International Common Petroleum Deposit*, 26 INT'L & COMP. L.Q. (1977) 324, 332 and 333. In my opinion, this conclusion is one proof of Onorato's traditional 'Grotian' research methodology.

29 *See* J.E. Horigan, *supra*, n. 19.

30 William T. Onorato, *Apportionment of an International Common Petroleum Deposit*, 26 INT'L & COMP. L.Q. (1977) 324, 332.

31 *Id.*, at 333.

32 *Id.*, at 336.

33 *See* Rainer Lagoni, *supra*, n.19, at 215–243.

Lagoni wanted to 'analyze the state practice [...] concerning common oil and gas deposits in order to determine the legal problems arising from those deposits and, in addition, to discover whether the solutions adopted may be considered a part of customary international law'.[34] Altogether, he treated four issues – two exposed directly in his primary analysis, one treated as a secondary matter and one residually: (i) the definition of the supreme authority over each part of a common deposit; (ii) the protection provided by international law to a state against infringements of its parts of a deposit; (iii) the relevance of such deposits to the delimitation of territorial or continental shelf areas; (iv) the 'real problems' which arise if and when a state or concessionaire begins to explore and exploit a common deposit, in particular in the context of a disparity in technical capacities; and (v) the rules that should be applied in areas of overlapping claims.[35] To complete this analysis, Lagoni divided the work into five part. It is important to note that he drew a sharp distinction between the legal criteria applicable to situations where a legal solution was ultimately achieved and the legal criteria applicable to situations where disputes and claims persisted.

The first part stated that the principles of territorial sovereignty, exclusive sovereign rights and territorial integrity applied, leading to state responsibility if breached.[36] In the transition between the first and second parts, Lagoni expressed that the main concerns in a common deposit of oil and gas arose from their border-straddling characteristics and from the technical impossibility of determining clear limits and sovereignty rights.[37] Accordingly, '[i]n the absence of an agreement on the exploration and exploitation of the common deposit', he asked, 'what are the rights and duties of the states concerned?'[38] Lagoni wrote that there was no consensus in international law or a 'pure' answer to this question, discarding four possible answers and expressing that they were 'just a trend'.[39] In the

---

34 *Id.*, at 215, 236, and 240.
35 *Id.*, at 215, 216, and 239.
36 *Id.*, at 216 and 217.
37 *Id.*, at 216–218.
38 *Id.*, at 219.
39 He understood that the prior application rule had become obsolete as a principle of civilized peoples, superseded by cooperative or unitised exploitation, and that it was wasteful and uneconomic. The rule of sovereignty over the subsoil to the common deposits of liquid minerals, indicated by Juraj Andrassy in 1951, was uncooperative and would lead to a 'silent war' of competitive drilling. He asserted

third part, he addressed the previous question with an overview of prac-tice-agreements for previously discovered common deposits and agree-ments including mineral deposit clauses for future discoveries,[40] which revealed in the former case that no form of joint property right or uniform pattern of systematic cooperation could be found,[41] and in the later that the general use of mineral deposit clauses[42] permitted the description of coop-eration as 'a firmly settled practice of those states whose interests [were] specially affected'.[43]

His fourth and fifth parts have a special importance to this study. The fourth part contained elements of causal inference methodology. To con-clude that there was a customary obligation supported by practice and *opinio iuris*, he understood that there was a need to negotiate in good faith if a common deposit of liquid minerals was discovered by any state.[44] He

---

that in 1979 there was no certainty about the obligation to inform and consult about common deposits, as proposed by Hugo Barberis. Finally, he rejected the idea that the recognition of joint property rights, as proposed by Onorato, could be regarded as a general state practice or general principle of law. *See* Rainer Lagoni, *supra*, n. 15, at 219–221.

40 Rainer Lagoni, *supra*, n. 15, at 222–233.

41 *Id.*, at 228.

42 A mineral deposit clause provides for cooperation should a common deposit of liq-uid minerals be discovered in the future. The first clause of this type was included in the Agreement between the Government of the United Kingdom of Great Britain and Northern Ireland and the Government of the Kingdom of Norway Relating to the Delimitation of the Continental Shelf Between the Two Countries (London 10 March 1965, 551 UNTS 214).

43 This type of clause has been used in thirty out of sixty boundary delimitation agreements since 1942 and in nearly all continental shelf delimitation agreements since 1970. Rainer Lagoni, *supra*, n. 15, at 233.

44 Lagoni at 235 *et seq*. The obligation to negotiate in good faith implies the duty to inform and consult the other state in case a deposit is discovered, and to refrain from unilateral exploitation before a process of negotiation. An agreement leads to the need to develop any joint exploitation using the best technical means available to a state taking into account factual circumstances, to define the delimitation of the deposit to determine its contents, and to define the standard of share (in pro-portion, equitable share, etc.). The customary obligation to negotiate, however, does not imply a need to achieve an agreement, but it certainly does provide for two important obligations: (i) to avoid damage to the other state and to provide indemnity in case of responsibility; and (ii) to avoid exploitation even if there is willful inaction from the other state, except where this is estopped or there is acquiescence from the inactive state. Nevertheless, ultimately, '[c]onsidering [...] that violation of territorial sovereignty is hard to prove in the case of common

asserted that there were causes (*independent variables*) that explained state practice: (i) existing 'soft law',[45] (ii) domestic politics and (iii) geopolitical interest.[46] At this point, it seemed that Rainer Lagoni was displaying hints of Rational Functionalism.[47]

In the fifth part of his work Lagoni returned to a Grotian perspective.[48] He stated that '[t]he customary duty to seek agreement on common deposits of liquid minerals in good faith always applie[d] to those states whose frontiers or dividing lines [were] duly established in the areas concerned'.[49] He concluded that there was no compulsory norm in the case of

---

deposits of liquid minerals, one can realistically assume that a breakdown of negotiations in the absence of a compulsory obligation for dispute settlement could easily lead to competitive drilling on both sides of the frontier or dividing line'. Rainer Lagoni, *supra*, n. 15, at 235–239.

45 In general, *see* Kenneth W. Abbott, Robert O. Keohane, Andrew Moravcsik, Anne-Marie Slaughter, and Duncan Snidal, The *Concept of Legalization,* in INTERNATIONAL LAW AND INTERNATIONAL RELATIONS, 115. *Also* Kenneth W. Abbot and Duncan Snidal, *Pathways to International Cooperation, in* THE IMPACT OF INTERNATIONAL LAW AND INTERNATIONAL COOPERATION, 50. Some of the United Nations General Assembly are: GA Res. 2995 (XXVII), 2996 (XXVII), and 3129(XXXVIII). *See* Rainer Lagoni, *supra*, n. 15, at 228 n. 86.

46 States need to maintain a position in the international arena, and a way of doing so is to demonstrate their ability to protect their territorial integrity and sovereign rights. *See* Rainer Lagoni, *supra*, n. 15, at 235.

47 For a general explanation of the elements and dynamics of Rational Functionalism, *see* Beth A. Simmons and Lisa L. Martin, *International Organizations and Institutions, in* INTERNATIONAL LAW AND INTERNATIONAL RELATIONS, 195.
   See James Fearon and Alexander Wendt, *Rationalism v. Constructivism: A Skeptical View, in* HANDBOOK OF INTERNATIONAL RELATIONS (2007) 52, 54.

48 Lagoni mentions that several states' '[u]nilateral declarations or national legislation on the continental shelf [...] have expressed their intention to delimit the shelf shared with another state by agreement in accordance with equitable principles'. Additionally, he underlines the economic interest of states, explaining that '[s]tates normally agree to cooperate with regard to a common deposit of liquid minerals in order to protect their own interests. These interests are primarily of an economic nature. If both states are going to exploit the deposit at the same time, they are interested in cutting extraction costs and achieving maximum production. If only one state intends to exploit the common deposit, there are also legal reasons for the other state to cooperate. Without cooperation, prejudicial exploitation would be unavoidable because of the physical conditions of any common deposit'. Rainer Lagoni, *supra*, n. 15, at 234.

49 *See* Rainer Lagoni, *supra*, n. 15, at 237.

persistent claims .[50] States were in any case free: (i) to accept or reject the principle of 'unity of the [liquid mineral] deposit' suggested by Onorato and by M. W. Mouton[51] when achieving a boundary agreement; (ii) to maintain their positions regarding the claim; or (iii) to achieve special regimes for the deposit.[52] Eventually, the entry into force of UNCLOS changed Lagoni's position and led him to find that as regards exploitation, there was an obligation of mutual restraint complemented by a duty to cooperate on states interested in a common deposit.[53]

As can be observed, this fifth part added some inferential elements to the previous ones: despite asserting that '[t]here is, of course, no conclusive catalog of relevant factors',[54] he points out that the historical relations between states and third party concerns and participation can influence the configuration of cooperation regimes.[55]

Second Stage

The second stage in the literature on offshore oil and gas cooperation was crafted by a group (*hereinafter*, 'The Group') of specialists from different disciplines who held a conference in Durham in July 1994. Since then, their work has addressed not only offshore oil and gas deposits, but all forms of natural resources.[56]

---

50 It is interesting that he does not carry over the obligation to cooperate exposed in Part Four as the leading principle in these cases as well.

51 Mouton, *The Continental Shelf*, 85 RECUEIL DES COURS (1954, 1) 347, 422.

52 Lagoni mentions coordinated exploitation, transboundary unitization, exemption of mineral resources in an agreement, and joint exploitation – which is the author's suggestion. *See* Rainer Lagoni, *supra*, n. 15, at 242.

53 *See* Rainer Lagoni, *Interim Measures pending Maritime Delimitation Agreements*, 78 Am. J. Int'l Law (1984) 345, 357.

54 Rainer Lagoni, *supra*, n. 15, at 241.

55 *Id.*

56 The main contributors in the area of offshore oil and gas cooperation were: Gerald Blake, Charles Robson, Rodman R. Bundy, Richard E. Swarbrick, Ian Townsend-Gault, William G. Stormont, David Ong, Jeremy Carver, Greg Englefield, Gideon Biger, Peter Odell, Paul Tempest, Victor Prescott, Mark Valencia, Jon M. Van Dyke, Daniel Dzurek, Sergei Vinogradov, Henry Huttenbach, William Dunlap, Bruce Blanche, Patrick Amstrong, Viv Forbes, Keith Highet, Ibrahim F. I. Shihata, William T. Onorato, Masahiro Miyoshi, D.H. Anderson, Albert Utton, Alberto

Though they retained some elements of Grotian methodology, The Group – led by Gerald H. Blake, Martin Pratt and Clive Schofield[57] – innovated with its multidisciplinary perspective, exploring an area previously marked by a narrow, possibly *microscopic*, view.

The context in the 1990s was that the UNCLOS – already exercising strong influence over international law and permanent sovereignty over natural resources – was fully in force, though slightly attenuated by environmental concerns.[58] From a purely legal perspective, the main topics The Group treated were: (i) the existence of an obligation to cooperate in good faith;[59] (ii) the forms that cooperation agreements could take;[60] (iii) the existence or non-existence of an obligation to adopt a specific form of

---

Szekely, Ulises Canchola, Carmen Pedrazzini, William Waggoner and Ross Shipman.

57 These three authors were the main editors of THE PEACEFUL MANAGEMENT OF TRANSBOUNDARY RESOURCES (Gerald H. Blake *et al.* eds., 1995) and BOUNDARIES AND ENERGY: PROBLEMS AND PROSPECTS (Gerald Blake *et al.* eds., 1998). See also, Churchill, R.R., '*Falkland Islands – Maritime Jurisdiction and Cooperative Arrangements with Argentina,*' *Current Legal Developments,* 46 International and Comparative Law Quarterly 463–477 (1997); Alberto Szekely, *The International Law of Submarine Transboundary Hydrocarbon Resources: Legal Limits to Behavior and Experiences for the Gulf of Mexico,* 26 Natural Resources Journal 766 (1986).

58 NICHO SCHRIJVER, SOVEREIGNTY OVER NATURAL RESOURCES 120 (1997).

59 *See* Charles Robson, *Transboundary Petroleum Reservoirs: Legal Issues and Solutions,* in THE PEACEFUL MANAGEMENT OF TRANSBOUNDARY RESOURCES 3, 3 (Gerald H. Blake *et al.* eds., 1995); Gerald H. Blake and R.E. Swarbrick, *Hydrocarbons and International Boundaries: A Global Overview,* in BOUNDARIES AND ENERGY: PROBLEMS AND PROSPECTS 3, 14 (Gerald Blake *et al.* eds., 1998); Ibrahim F. I. Shihata and William T. Onorato, *Joint Development of International Petroleum Resources in Undefined and Disputed Areas,* in BOUNDARIES AND ENERGY: PROBLEMS AND PROSPECTS 433, 433(Gerald Blake *et al.* eds., 1998); Masahiro Miyoshi, *International Maritime Boundaries and Joint Development: A quest for a Multilateral Approach,* in BOUNDARIES AND ENERGY: PROBLEMS AND PROSPECTS 453, 453 (Gerald Blake *et al.* eds., 1998).

60 *See* Robson, *supra* note 63, at 4, 6, 9, and 12; *Also* D.H. Anderson, *Strategies for Dispute Resolution: Negotiating Joint Agreements,* in BOUNDARIES AND ENERGY: PROBLEMS AND PROSPECTS 473 (Gerald Blake *et al.* eds., 1998).

cooperation agreement;[61] and (iv) the existence of an obligation of mutual restraint where cooperation was impossible to achieve.[62]

As already mentioned, from a theoretical perspective, The Group did not abandon its Grotian roots.[63] Nevertheless, it innovated, and in the end their perspective was consolidated into Rational Functionalism.[64] In this sense, they introduced many elements into the analysis of regime formation:

- Different levels of interaction in the international arena were recognised:

  - '…boundaries will become highly permeable. One of the ways in which boundaries are changing from being "hard" to "soft" is in relation to resource management'.[65]
  - 'At the domestic level, then, ocean management requires a degree of functional integration between the maritime sectors, but in such a way that each area of activity is able to proceed, but without danger of adverse impacts on the others'.[66]

- Different variables become essential to the configuration of Political Will,[67] which precedes foreign policy:

---

61 Rodman R. Bundy, *Natural Resource Development (Oil and Gas) and Boundary Disputes, in* THE PEACEFUL MANAGEMENT OF TRANSBOUNDARY RESOURCES 23 (Gerald H. Blake *et al.* eds., 1995).

62 David M. Ong, *Joint Development of Common Offshore Oil and Gas Deposits: "Mere" State Practice or Customary International Law?*, 93 Am. J. INT'L L.771, 798. Also, Miyoshi the basic at 14 note 65.

63 'States rarely enter into boundary agreements for abstract reasons. They do so because they need to have the boundary determined, usually for the purpose of resource planning and management. It is no accident that the majority of ocean boundaries in the world today involve important petroleum producing countries', Ian Townsend-Gault and William Stormont, *Offshore Petroleum Joint Development Arrangements: Functional Instrument? Compromise? Obligation?, in* THE PEACEFUL MANAGEMENT OF TRANSBOUNDARY RESOURCES 51, 58 (Gerald H. Blake *et al.* eds., 1995). *See also* Ong, *supra* note 66, at 795.

64 *See* Townsend-Gault and Stormont, *id.*, at 63. *Also* Peter R. Odell, *Hydrocarbons: The Pace Quickens, in* 28, 37 BOUNDARIES AND ENERGY: PROBLEMS AND PROSPECTS 453, 453 (Gerald Blake *et al.* eds., 1998).

65 Gerald Blake, *Introduction, in* THE PEACEFUL MANAGEMENT OF TRANSBOUNDARY RESOURCES xiii, xvii (Gerald H. Blake *et al.* eds., 1995).

66 Townsend-Gault and Stormont, *supra* note 67, at 60.

67 *Id.*, at 61.

- International politics at global level[68] and at a regional level.[69]
- The global economy:[70] in particular the role and risks of oil market.[71]
- The context provided by existent rules of Public International Law.[72]
- The interest of third parties.[73]
- Geological characteristics and existent technical resources.[74]
- The historical context.[75]

Finally, the empirical effects of legal practice were taken into account,[76] which ultimately provided the analysis with a dynamic substance.[77] By the

---

68 'Law of the sea developments up to the mid-1960s tended to emphasize unilateral rights and the phraseology of the doctrine of the continental shelf, with its emphasis on "sovereign rights for the purposes of exploration and exploitation," tends to emphasize this. But this concept must now be read together with that of interdependence' *id.*, at 59.

69 'ocean management requires a degree of functional integration between the maritime sectors, but in such a way that each area of activity is able to proceed, but without danger of adverse impacts on the others. [T]his process must be repeated at the sub-regional, regional, and then inter-regional level' *id.*, at 60. *Also*, Masahiro Miyoshi, *supra* note 66, at 454.

70 *See* Townsend-Gault and Stormont, *supra* note 67, at 61.

71 *See* Odell, *supra* note 68, at 28. *Also*, Ibrahim F.I. Shihata and William T. Onorato, *supra* note 63.

72 *See* Townsend-Gault and Stormont, *supra* note 67, at 59. *Also* Keith Highet, *New Courts and Old, Old Law and New, and Problems to Come*, in BOUNDARIES AND ENERGY: PROBLEMS AND PROSPECTS 415 (Gerald Blake *et al.* eds., 1998).

73 *See* Masahiro Miyoshi, *supra* note 66. *Also* Keith Highet, *id.*, at 420.

74 *See* Richard E. Swarbrik, *Oil and Gas Reservoirs Across Ownership Boundaries: The Technical Basis for Apportioning Reserves, in* THE PEACEFUL MANAGEMENT OF TRANSBOUNDARY RESOURCES 41, 47 (Gerald H. Blake *et al.* eds., 1995).

75 *See* Townsend-Gault and Stormont, *supra* note 67, at 62.

76 Analysing the large oilfield on the Iran–Abu Dhabi border (the Sassan or Abu al Bukoosh (ABK) field), Bundy stated that 'In practice, only a modest amount of co-ordination was achieved. Prior to the Iranian Revolution, a joint committee was established primarily as a conduit for the exchange of technical information relating to the field, but this line of communication broke down after the Revolution and remained so during the Iran-Iraq war'. Bundy, *supra* note 65, at 26. *Also* Robson, *supra* note 63, at 9.

77 Townsend-Gault and Stormont, *supra* note 67, at 51. *Also* Ong, supra note 66, at 795.

end of the 1990s, the issue had become 'whether a suitable international legal framework [was] evolving that would enable states to efficiently exploit shared mineral resources at present or in the near future.'[78]

Third Stage

The first decade of the twenty-first century found two academic groups taking different paths. One seemed to be returning to the original Grotian perspective, while the other was developing its literature under the influence of human-rights/environmentally-concerned/global-governance thought.[79]

The first group of academics focused primarily on the enumeration and description of the existing bilateral agreements on offshore cooperation.[80] Only secondarily did they mention any of the elements which had been relevant to the negotiation and regime configuration.[81] They did con-

---

78 Some authors considered the answer to be that 'unitization [appeared] to be the legally appropriate and efficient solution for exploiting the common deposit', even if joint development was the preferred alternative. *See* Ong, *supra* note 66, at 780 and 800.

79 *See* HUMAN RIGHTS IN NATURAL RESOURCE DEVELOPMENT (Zillman el al. eds.); *also* THE NEW ENERGY PARADIGM (Helm *et al.* eds., 2007).

80 The works by this group include: OCEAN DEVELOPMENT AND INTERNA-TIONAL LAW (2005); INTERNATIONAL BOUNDARIES RESEARCH UNITS, MARITIME BRIEFING (2001); Schofield, C.H., *Unlocking the Seabed Resources of the Gulf of Thailand*, 29 Contemporary Southeast Asia 286–308 (August 2007);); Richard J. McLaughlin, *Hydrocarbon Development in the Ultra-Deepwa-ter Boundary Region of the Gulf of Mexico: Time to Reexamine a Comprehensive U.S.–Mexico Cooperation Agreement*, 39 Ocean Development and International Law 1–31 (2008); Clive Schofield, *Blurring the Lines? Maritime Joint Develop-ment and the Cooperative Management of Ocean Resources, in* FRONTIER ISSUES IN OCEAN LAW: MARINE RESOURCES, MARITIME BOUND-ARIES, AND THE LAW OF THE SEA (The Berkeley Electronic Press, 2009). *Also* Luis E. Rodriguez-Rivera, *Joint Development Zones and other cooperative Management Efforts Related to Transboundary Maritime Resources: A Caribbean and Latin American Model for Peaceful Resolution of Maritime Boundary Dis-putes, in* FRONTIER ISSUES IN OCEAN LAW: MARINE RESOURCES, MAR-ITIME BOUNDARIES, AND THE LAW OF THE SEA (The Berkeley Electronic Press, 2009).

81 The elements considered by authors are mostly: political will, normative context, the history of bilateral relationships and third party conduct (third states, oil and gas companies and NGOs).

tribute, nevertheless, by adding some elements to those already considered by previous authors: (i) the geographical context of the parties to an agreement;[82] (ii) the character or 'strength' of the agreement – 'soft' law or 'hard' law;[83] and (iii) the identity of the states involved in the agreement.[84]

The second group of academics were Social Constructivists.[85] Their perspective can be summarised in the work of Eyal Benvenisti.[86] Benvenisti started a project to examine what guidelines international law offered for the resolution of the conflict over the use of the shared freshwater resources in the Middle East. To him, 'increasing scarcity on the one hand, and negotiations towards a peaceful settlement of the Arab-Israeli conflict on the other, lent a sense of urgency even to such an academic exercise'. Soon he discovered that the basic principles of international law on shared fresh water were being disputed, including by the governments in the Middle East, and that therefore negotiators were not being significantly constrained by the law. Trying to explain why the law failed to offer shared norms, he expanded his initial inquiry. As he said, he 'ventured into new disciplines. The study [...] taught [him] that [he] should focus on ecosystems rather than on watercourses. The study of game theory and public choice theory permitted [him] to explore the roots of the failure of collective action in the national, regional, and international arenas'.[87]

Benvenisti's work can be divided into two phases. The first phase explained state conduct towards neighbours by analysing the conditions for collective action in the management of transboundary resources exposed by the preferences expressed in the transnational competition over resources shaped by states.[88] Observation of state practice led him to

---

82 Schofield, *supra* note 84, at 6.
83 *Id.*
84 Rodriguez-Rivera, *supra* note 84, at 11.
85 *See* Beth A. Simmons and Lisa L. Martin, *supra*, note 51, at 198.
86 EYAL BENVENISTI, SHARING TRANSBOUNDARY RESOURCES. INTERNATIONAL LAW AND OPTIMAL RESOURCE USE (2002).
87 *See* Benvenisti, *supra*, note 90, at xi.
88 'In exploring the potential for collective state action, this book breaks away from the distinction between domestic and international processes. The systemic failures of states that derive from their heterogeneity in their management of natural resources affect similarly domestic and international resources. Domestic conflicts of interest – between farmers and city dwellers, between producers and consumers – are responsible for distorted governmental choices, whether on the domestic or

conclude that states failed to act together with their neighbours to protect their shared resources, and that those failures could be corrected.[89] The second phase work examined what responses international law had to offer to induce states to choose to cooperate. He concluded that the divergence within contemporary international law, particularly the clash between two conflicting approaches – the 'philosophy of disengagement' and 'the philosophy of integration'[90] – requires a final insight: international tribunals, particularly the International Court of Justice (ICJ), are authorised to develop international law in accordance with the goal of effi-

---

the international level. In fact, many domestic interest groups cooperate, with foreign interest groups in order to impose their externalities on their respective rival domestic groups. The better-organized and, hence, more politically effective domestic interest groups – usually producers, employers, and service-suppliers – cooperate with their counterparts in different states to exploit collectively the less organized groups in those in transboundary ecosystem institutions in ways that will ensure informed and unbiased decisions. Such institutions could sustain markets for some of the uses of the resource, say, for trade in pollution permits, for water for irrigation, or for reclaimed sewage water intended for agriculture, provided the institution can ensure the attainment of the tasks of resource, claims, and risk-management and can comply with the external normative constraints'. *Id.*, at 19.

89 *Id.*, at 233 and 234.
90 As we trace the evolution of the relevant norms of international law, we will notice a clash between two philosophies. The first philosophy – what I will term 'the philosophy of disengagement' – characterises the attitude of many governments as well as of the World Bank, whereas the second – 'the philosophy of integration' – characterizes the approach advocated mainly by scientists and scholars, but particularly by judges in international litigation. The two contradictory philosophies pushed the codification efforts of international ecosystem law throughout the twentieth century in two opposite directions. Whereas disengagement philosophy strives to limit common ownership among riparian states to the lowest possible minimum, the philosophy of integration suggests that common ownership and inclusive management is not only an inescapable outcome, but also a beneficial one. The disengagers look to international law as a system of rules that could minimise friction among riparians and resolve interstate disputes through adjudication and rigid arm's-length agreements assigning the allocation of rights and obligations as clearly as possible. In contrast, the integratives seek more, rather than less, friction as the preferred alternative and thus opt for the management of disputes through negotiation leading to flexible agreements which establish joint management institutions. The integratives acknowledge that states are no longer 'bound by international law only at their own discretion [but instead have] the responsibility to develop and implement international law in order to further the interests of humanity'. *Id.*, at 157.

ciency. The first phase of Benvenisti's work showed causal inference methodology. The second phase of Benvenisti's work was shaped by his own conjectures and suggestions:

> The question is whether the integrative philosophy will prevail over the disengagement philosophy in the long run. Some even worry about the growing propensity of some states to resort to force to secure shares in key transboundary resources such as fresh water. I belong to the camp of the cautious optimists who are inclined to adopt the view that as demands for fresh water and the environment intensify and diversify and supplies dwindle, interdependency becomes greater. I believe that transnational collective action in the utilization of transboundary resources can, in principle, provide optimal and sustainable results. A bleak future of wars over control of water resources is not an unavoidable tragedy in our new millennium. Growing interdependency increases the incentive of neighboring states to cooperate in formulating entitlements and norms and in managing shared resources. Thus, in the near future, as supplies of finite trans-boundary resources in general fail to meet rising demands, more and more states may be unable to afford the luxury of ensuring their relative power gap through unilateral appropriation. They will sooner or later opt for making the most from the collective-action situation'.[91]

Codification: The Work of the International Law Commission (2000 – 2010)

The United Nations International Law Commission (the "Commission"), at its fifty-fourth session (2002), decided to include the topic "Shared natural resources" in its programme of work and appointed Mr. Chusei Yamada as Special Rapporteur. A Working Group was also established to assist the Special Rapporteur in sketching out the general orientation of the topic in the light of the syllabus prepared in 2000. The Special Rapporteur indicated his intention to deal with confined transboundary groundwaters, oil and gas in the context of the topic and proposed a step-by-step approach beginning with groundwaters. From its fifty-fifth (2003) to its sixtieth (2008) sessions, the Commission received and considered five reports from the Special Rapporteur. During this period, the Commission also established four working groups, the first of which was chaired by the Special Rapporteur and the other three by Mr. Enrique Candioti. The first working group, established in 2004, assisted in furthering the Commission's consideration of the topic. The second working group, established

---

91 *Id.*, at 234 and 235.

in 2005, reviewed and revised the 25 draft articles on the law of trans-boundary aquifers proposed by the Special Rapporteur in his third report (A/CN.4/551 and Corr.1 and Add.1) taking into account the debate in the Commission. The third working group, established in 2006, completed the review and revision of the draft articles submitted by the Special Rapporteur in his third report, culminating in the completion, on first reading, of the draft articles on the law of transboundary aquifers (2006). The fourth working group, established in 2007, assisted the

Special Rapporteur in considering a future work programme, in particular the relationship between aquifers and any future consideration of oil and gas, consequently agreeing with the proposal of the Special Rapporteur that the Commission should proceed to a second reading of the draft articles on the law of transboundary aquifers in 2008 and treat that subject independently of any future work by the Commission on oil and gas. Moreover, the Commission, at its fifty-eighth session (2006), adopted, on first reading, draft articles on the law of transboundary aquifers consisting of 19 draft articles, together with commentaries thereto. The Commission, at its sixtieth session (2008), adopted, on second reading, a preamble and a set of 19 draft articles on the law of transboundary aquifers, with a recommendation that the General Assembly: (a) take note of the draft articles and annex them to its resolution; (b) recommend to States concerned to make appropriate bilateral or regional arrangements for the proper management of their transboundary aquifers on the basis of the principles enunciated in the draft articles; and (c) consider, at a later stage, and in view of the importance of the topic, the elaboration of a convention on the basis of the draft articles.

The Working Group held one meeting on 3 June 2009 and exchanged views on the feasibility of any future work by the Commission on the issue of transboundary oil and gas resources. It addressed several aspects, including whether there was a practical need for work on oil and gas; the sensitivity of the issues to be addressed; the relationship between the issue of transboundary oil and gas resources and the question of boundary delimitations, including maritime boundaries; and the difficulties in the collection of information relating to the practice in this field. Some members, while recognizing the specificities of each situation involving the exploration or exploitation of transboundary oil and gas resources, were of the view that there might be a need to clarify certain general legal aspects, in particular in the field of cooperation. Several members emphasized the need for the Commission to proceed cautiously with regard to oil and gas, and to be responsive to the views expressed by States. Some members pointed to the fact that the majority of Governments who expressed themselves on this issue did not favour future work by the Commission on oil and gas, or expressed reservations thereto. However, the point was also made that the number of written responses received so far, although substantial, was still insufficient for the Commission to make an assessment on whether it should undertake any work on this subject. The view was expressed that the General Assembly had already considered that oil and gas were going to be part of the topic "Shared natural resources." He recalled that the Commission had first dealt with the problem of shared natural resources during its codification of the law of non-navigational uses of international watercourses. At the time, the Commission had decided to exclude confined groundwaters unrelated to surface waters from the topic; nonetheless, it was also felt then that a separate study was warranted due to the importance of confined groundwaters in many parts of the world. It was noted that the law relating to groundwaters was more akin to that governing the exploitation of oil and gas. Under the topic, the Special Rapporteur proposed to cover confined transboundary groundwaters, oil and gas and to begin with confined transboundary groundwaters. In order to ascertain the extent to which the principles embodied in the 1997 Convention on the Law of the Non-navigational Uses of International Watercourses could be applicable, he deemed it indispensable to know exactly what such groundwaters were. He also pointed out that the work carried out on the topic of international liability, particularly regarding the prevention aspect, would be relevant. The Special Rapporteur concluded by indicating that he intended to conduct studies on the practice of States with respect to uses and management, including pollution prevention, and cases of conflicts, as well as domestic and international rules. Furthermore, he would attempt to extract some legal norms from existing regimes and possibly prepare some draft articles. The view was expressed that a single report encompassing oil and gas in addition to groundwater would have given a better overview of the subject, particularly as regards the principles applicable to the three resources and the differences among them. It was suggested that priority be given to the subject of confined groundwaters and, in particular, to the issue of non-connected groundwater pollution. The view was expressed that any consideration of the topic of oil and gas should be postponed until the Commission has concluded its work on groundwaters. The working paper noted that a majority of States was of the view that the transboundary oil and gas issues were essentially bilateral in nature, as well as highly political and technical, involving diverse situations. Doubts were expressed as to the need for the Commission to proceed with any codification exercise on the issue, including the development of universal rules. It was feared that an attempt at generalization would inadvertently lead to additional complexity in an area that may have been adequately addressed through bilateral efforts. Given that oil and gas reserves were often located on the continental shelf, there was also a concern that the subject had a bearing on maritime delimitation issues. Maritime delimitation, which, in political terms, was a very delicate issue for the States, would be a prerequisite for the consideration of this as subtopic, unless the parties had mutually agreed not to deal with delimitation. Furthermore, it was considered that the option of collecting and analysing information about State practice concerning transboundary oil and gas or elaborating a model agreement on the subject would not lead to a fruitful exercise for the Commission, precisely because of the specificities of each case involving oil and gas.

The sensitive nature of certain relevant cases could well be expected to hamper any attempt at a sufficiently comprehensive and useful analysis of the issues involved.

Lastly, the Report of the International Law Commission on the work of its sixty-second session (3 May to 4 June, and 5 July to 6 August 2010) decided once more to establish a Working Group on Shared natural resources, chaired by Mr. Enrique Candioti. The Working Group had before it a working paper on oil and gas (A/CN.4/621) prepared by Mr. Shinya Murase. At its 3069th meeting, on 27 July 2010, the Commission took note of the oral report of the Chairman of the Working Group on Shared natural resources and endorsed the recommendation of the Working Group. The Working Group considered all aspects of the matter taking into account the views of Governments, including as reflected in the working paper by Mr. Murase. In light of the foregoing, it decided to recommend that the Commission should not take up the consideration of the transboundary oil and gas aspects of the topic "Shared natural resources".

3 The Origin of this study

My interest in the study of oil and gas cooperation regimes for controversial sea areas started with the idea that they could serve better than other traditional methods for the settlement of existing and potential disputes. However, I understood that even though negotiation and the conclusion of agreements was an exceptional achievement, it was only a point of departure. My understanding was that everything which came afterwards was the really important issue. To be precise, I found that after 1994 Durham Conference, Gerald Blake had expressed regret at the lack of systematic literature with a dynamic-epistemological approach to offshore cooperation regimes.[92] That provided the decisive impulse for me to commit to this research.

Over the last decade, as Amitav Acharya and Alastair Iain Johnston have noted, the study of international institutions has been more interested in whether institutions matter than on how they matter or actually work. They engaged in that line of thought and forged a model of comparative analysis which added sociological elements to a research design conceived previously by Barbara Koremenos, Charles Lipson and Duncal Snidal,[93] as they considered that their perspective was too rationalist and could omit elements found in regime dynamics.[94] They thought that combining rationalist literature with other methodology more common in sociological approaches 'allow[ed] for a fuller test of the different ways in which institutions may affect efficacy and the nature of cooperation, and [struck] a balance between conceptual tractability on the one hand and a more complete test of a fuller range of plausible hypotheses'.[95]

---

92 In this sense, 'Martti Koskenniemi dichotomizes all of international legal argumentation into a debate between the apologists and the utopians – those who accept that international law reflects whatever states do and those who would have international law transcend and constrain state behavior'. Anne-Marie Slaughter, *supra* note 18, at 26.

93 THE RATIONAL DESIGN OF INTERNATIONAL INSTITUTIONS (Koremenos *et al.* eds., 2004).

94 *See* CRAFTING COOPERATION 12 (Acharya and Johnston eds., 2007).

95 *Id.*, at 22.

In my opinion, Acharya and Johnston's perspective was correct. In general terms, their research design also suited my project, despite being originally intended to be applied to regional institutions and not to bilateral agreements. From my point of view, bilateral agreements constitute regimes; 'orthodox' as it might be, there is always the possibility for a treaty to achieve wider effectiveness and acquire some degree of 'independence' from its parties.[96] Consequently, I decided to keep Acharya and Johnston's research structure, with some of its variables reformulated. Finally, I added variables that I consider relevant to the topic under analysis.

Methodology: Design and structure of this study

The first objective in this study is to identify and explain the design principles of a selection of offshore cooperation regimes. The second objective is to analyse the subsequent dynamics –the nature of cooperation – generated by the regime's design.[97] By *design*, I mean those 'formal and informal rules and organizational features that constitute the cooperation institution and that function as either the constraints on actor choice or the bare bones of the social environment within which agents interact, or both'.[98]

Variables

To achieve my aforementioned aims, I brought together a number of complex cases and applied a two-part analysis methodology to them. In the first part I treat cooperation agreements as a dependent variable by establishing an array of possible explanatory reasons (which explain an element of an agreement without necessarily being its immediate cause) explored by their design. In the second part I treat the agreements' regimes as an independent variable and observed their performance since formation. The

---

96 *See* Stephen D. Krasner, *Structural Causes and Regime Consequences: Regimes as Intervening Variables, in* INTERNATIONAL LAW AND INTERNATIONAL RELATIONS 3 (Beth A. Simmons and Richard H. Steinberg eds. 2006).

97 *See* Finnemore, Martha and Sikkink, Kathryn, *International Norm Dynamics and Political Change*, 52 International Organizations 887–917 (1998).

98 *See* Aracharya and Johnston, *supra* note 94, at 15.

methodology as a whole can be described analogically, using the language of complex adaptive systems: a range of complex cases are brought together, general parameters are provided for them and they are observed to see what kind of intellectual 'emergent property' this yields as a plausible conclusion about the evolution and effect of institutions in their contexts.[99]

It is important to recall that this work attempts to avoid any assumptions about the superiority or otherwise of any given theoretical or ontological orientation, to enable the search for plausible explanations and to permit full sight of the complexity behind the regime of the sea, on the one hand, and oil and gas issues on the other. As a consequence, I chose a method which combines the possibility of deductive and inductive hypotheses in testing and generation.

Type of cooperation problem. As a first cut, the origins and design features of the cooperation institution are looked at from a rational functionalist perspective: the form of an agreement will, in general, reflect the nature of the cooperation problem.[100]

Ideology and identity. Institutional designs can be affected by the ideology of the entrepreneurs.[101] Additionally, identity – here not just a function of common cultural features, (such as linguistic, racial, religious etc.) among states, but of shared norms, social purposes, cognitive models and

---

99 *Id.*, at 16.

100 '[A]ssuming rationality and some degree of shared understandings (common knowledge) about the preferences, beliefs, and strategies of other actors, this approach expects that there are optimal institutional designs for the cooperation problem at hand. This, for N-person prisoner's dilemma problems (where the issue is how to prevent actors from opportunistically defecting against each other), an institutions that can credibly monitor behavior, detect defection, and punish defectors will reduce the incentives to defect. For an assurance game, the cooperation problem is that, while all actors have a dominant cooperation strategy, they can't be sure that the others share this preference for a cooperative outcome. An institution that can provide information about the preferences and actions of all players will ensure that the Pareto-optimal outcome (mutual cooperation) is stable. Basically, it does not need monitoring or sanctioning mechanisms'. *Id.*, at 16.

101 In this sense, liberal ideology is supposed to encourage 'dense' institutions to become more 'permeable' to third party participation in bilateral affairs, and to allow the weakening of their traditional notions of sovereignty to a certain extent. Authoritarian or semi-authoritarian states do not permit intrusion or any form of interference in internal affairs. The same stance is often encountered with postcolonial ideologies.

the views of outgroups forged through political and economic interactions among culturally diverse units – could affect the willingness to negotiate, the process of negotiation, the approval of the agreement achieved, and forge the norms of that agreement.[102]

Systemic and subsystemic power distributions. This work opens by acknowledging that energy security has become an issue important enough to test traditional realist arguments about how institutional designs will reflect the interests of the most powerful states.[103]

General international normative and economic landscape. The general characteristics of hard and soft international law, norms, the state of the economy (global and regional), and finally the specific energy market context can all affect the actors' preferences when engaging in negotiations and concluding an agreement.

Domestic politics. A number of elements which impact on foreign policy and consequently on institutional design can be envisaged. Governments must deal with political groups, press and media, public opinion, and national companies when crafting their policies.[104] Additionally, pol-

---

102 This variable can have some influence on negotiations and agreement characteristics between states (between Muslim and non-Muslim countries; or between European, Latin American, African, Asian etc. countries).

103 Regionally or globally dominant countries can affect the treatment of the issues in question by promoting some form of agreement or the inclusion of particular clauses, as well as by choosing the development method to be incorporated in the agreement.

104 On the particular issue of domestic agents, Slaughter mentions that 'literature highlights three cutting edge issues for both international lawyers and political scientists, each of which supports the claim that domestic politics are as important for international lawyers to understand and integrate into their work as international politics:

(1) First, one of the most promising pathways for enhancing the effectiveness of an international legal regime is by bolstering or even triggering domestic political activity.

(2) Second, international law is made by states, but state positions do not spring fully formed from chancelleries or foreign ministries. Different social and governmental actors who actually succeed in being represented at the state policy-making level are the real sources of international law. Thus international law-making is better understood as a "bottom up" than a "top down" process.

(3) Third, the state itself must be reconceptualized as a two-level entity, a set of interaction between actors in domestic and transnational society and a wide

icies have their own evolution and local interaction.[105] In both cases, the variable can have an stimulatory or restraining influence on a state's foreign affairs.[106]

External institutions and non-state actors. Third parties, both public (states, regional organisations, international organisations, etc.) and private (multinational companies, NGOs), can provide organisational templates for newer institutions or reject prospective ones.

Technical characteristics: industrial, geographical, and geological. In the case of the development of offshore natural resources, all three of these elements are crucial for a state when balancing the contractual terms to be agreed.[107]

History. Finally, history can affect an institution's existing features. History is referred in two senses here. The first is history as manifested in historical memory. That is, how has the internalisation of appropriate institutional forms across time constrained the design options which current agents believe are available to them? The second is history as path dependence. How have increased returns to historical institutional features and the mechanisms for locking-in these features affected the current design of bilateral institutions? Normative and institutional path dependence could thus result from the transfer of the institutional features of previous institutions and from *bon voisinage* diplomacy.[108] Alternatively, the concept can come into play when the same continuing institution considers revisions to

---

array of government institutions'. *See* Anne-Marie Slaughter, *supra* note 1, at 47 and 48.

105 Domestic policies are understood in a general sense (e.g., the national economic model) or as they are formulated on a particular issue or area (regulatory laws, tax laws, energy and hydrocarbon industry laws, foreign investment regimes, financial regimes etc.).

106 States with negative policy environments, references or precedents are supposed to limit prospective investment, which is essential for high-tech industry.

107 Under-developed countries are assumed to be more likely to accept unbalanced terms in agreements, as they probably do not have the capacity to develop offshore natural resources. Additionally, financial aid has a contractual cost.

108 *See* Alan K. Henrikson, *Facing across Borders: The Diplomacy of Bon Voisinage*, 21 International Political Science Review 121–147 (Apr., 2000). *Also* Paul F. Diehl, Charlotte Ku and Daniel Zamora, *The Dynamics of International Law: The Interaction of Normative and Operating Systems, in* INTERNATIONAL LAW AND INTERNATIONAL RELATIONS, 426 (Beth A. Simmons and Richard H. Steinberg eds., 2006.).

its existing design features, either due to internal demands from one state or external pressures from new global events and norms.

The *dependent variable* for these independent variables is the *design of offshore institutionalised bilateral cooperation*: that is, the formal and informal rules and relationships that constitute the institution itself. Consequently, this work identifies four major features, modifying the original conceptions of Aracharya and Johnston.[109] *Scope* refers to the range of issues that the institution is designed to handle. The scope could be narrow, broad, intrusive, non-intrusive, detailed or sketchy. *Formal rules* refers to the explicit and 'legalised' regulations governing how decisions are made.[110] These relate both to an agreement's decision processes and to the management of hydrocarbon resources.[111] *Norms* refers to an institution's formal and informal ideology. What normative and causal arguments does the institution intend to promote? What normative and causal claims does it actually promote? Finally, *mandate* refers to the overall purpose of the institution. Is it designed to take immediate measures and normatively capable of this (operative), or is it just a forum to discuss potential solutions (programmatic)?

The objective of the second part of the study is to see how an agreement's design can affect cooperation. It does not intend to qualify this or to provide its sole description.[112] Accordingly, for an institutional design

---

109 *See* Aracharya and Johnston, *supra* note 94, 21, nn. 52–54.

110 *See* Duncan Snidal *et al.*, *supra* note 45.

111 To simplify analysis, I have adopted P. Carver's oil and gas development taxonomy: (i) geological cooperation, based on agreed rates of production by both parties working their own sectors of the deposit, the apportionment of production quotas being based on a knowledge of reserves determined by a joint commission; (ii) joint operations, which allows for equal sharing of production irrespective of the side of the border on which it occurs, with close operation between producers on both sides clearly being essential; (iii) unitised exploitation, where the development of a common deposit is driven by a single operator acting on behalf of the parties; and (iv)joint powers, where a common zone is established beyond a water depth of 1,000 metres in which both parties enjoy equal sovereign rights. *See* G.H. Blake and R.E. Swarbrick, *Hydrocarbons and International Boundaries: A Global Overview, in* BOUNDARIES AND ENERGY: PROBLEMS AND PROSPECTS 3, 14 (Gerald Blake *et al.* eds., 1998). For a purely legal description of hydrocarbon development regimes and contract forms, *see* BERNARD TAVERNE, PETROLEUM, INDUSTY AND GOVERNMENTS (1999).

112 As Acharya and Johnston recalled, 'quality' is measured by at least three dimensions: the degree of change that cooperation required in a state's original policies;

to be taken into account, it must have an effect on the cooperation dynamics or it will be disregarded. Where a design is of interest in this respect, I will explore ideas about how well it does *vis-à-vis* other plausible explanations.

These design features are important to deriving different, sometimes competing, hypotheses about the nature of cooperation. According to the traditional position of international lawyers, agreements are ultimately governed by their parties. Any sort of inclusion of third parties must be secondary and non-binding if the bilateral agreement is to be complied with correctly and without bias. However, social influence theory suggests that permeable institutions improve their status and effectiveness by including actors which might be even more important to the cooperation than the states parties themselves. Different types of treaty and formal resource management decision rules can also have different implications for cooperation. Bilateral agreements coping with shared natural resources are not based only on unanimity/consensus; the question is which rules expand and which dilute or obstruct the institution by holding decision-making hostage to the states' fluctuating political (good) will. Additionally, decisions are sometimes taken by representatives who might not have a complete understanding of oil and gas matters. Delegation to a more technical organ could offer greater substance, increase credibility with third parties and consequently strengthen commitment. The normative content of an institution could be important from both a rationalist and a socialisation perspective. It can mobilise domestic interests to try to capture national policy towards a cooperation institution. Alternatively, the normative content could be diffused into national policy processes through the socialisation of national representatives in the institution. As for its mandate, given that the subject is highly technical and the issue of great economic and geopolitical importance, national representatives can be expected to try to give primacy to the defence of national interests over

---

the degree to which cooperation affected the relative power of the state; and whether the state's cooperation was elicited through either positive or negative economic and social sanctions, or through normative acceptance. However, measuring 'quality' in this way also poses the risk of allowing an implicit normative bias to affect the measurement of the dependent variable. High quality cooperation can be interpreted as inherently more desirable. See Acharya and Johnston (eds.), *supra* note 2, at 264. *Also* George W. Downs, David M. Rocke, and Peter N. Barsoom, *Is the Good News About Compliance Good News About Cooperation?*, *in* INTERNATIONAL LAW AND INTERNATIONAL RELATIONS, 92.

any sort of additional emblematic aim the institution may forge. These and other hypotheses are used in the second part of each case analysis to examine the second dependent variable – *the nature of cooperation.*

The *nature of cooperation* is disaggregated into different components which might or might not be strongly present in each case, since these indicators are drawn eclectically from different streams of research on institutions. From constructivism is borrowed the idea of the *degree of normative and preference change.* To what extent does institutional design constitute a social environment within which actors are socialised to internalise new preferences, norms and roles? From rationalist institutional theory is borrowed the notion of the *degree of policy convergence across actors.* In addition, the *degree of institutionalisation and legalisation* is considered, since greater institutionalisation and legalisation suggests a deepening of cooperation. Another possible component concerns the *different routes to the above changes*: other than material interests, do persuasion and/or social influence play a role, depending on variation in the institutional design? A fifth component concerns the *degree of adjustment of prior policies and behaviours* which parties have to undergo when applying/performing an institution's cooperation plan? Does it require a great deal of change or very little? This speaks to the question of how radical the effect of institutional design is on actor practices. A sixth possible component concerns the *degree to which the institution (or the agents active in it) achieves its goals.* Does the arrangement actually produce the cooperative outcomes envisioned by its parties and participants? Finally, *to what extent does cooperation elicited by the institution have an impact on the 'problem' writ large?* The institution may meet its specific cooperative goals but these goals may have little overall impact on the difficulties with relationship or similar problems which nations face.

Endogeneity

Before enunciating the cases to be studied, it is important briefly to pause and treat the analytical problem of endogeneity.[113] Endogeneity could take two forms here: (i) the theoretical problem that states tend to support institutions which balance ease in establishment with preferred outcomes; and

---

113   *See* King, Keohane and Verba, *supra* note 14, at 184.

(ii) the strong interactive effect between design and efficacy. In its first form, endogeneity can be approached by looking for additional observable implications, where state preferences do not explain the design features over time.[114] Addressing its second form, treaties can be assumed to be difficult to negotiate and conclude and therefore institutional designs are sticky. In addition, it is easier to analyse institutions from their inception, and when considering their evolution, it is possible to drive sequential observations which co-opt only design-cooperation interactions.

Case Study

As it's been posed in the foreword, the number of cases for study is high and increasing. Unfortunately, time and space were a limit to this work, so the choice of cases had to be based on seeking geographical, temporal, social and institutional diversity. Consequently, the case study choices were reduced to the Middle East, America and Africa.

---

114 These observations include higher levels of cooperation than expected; high institutional autonomy and stronger effects over cooperation dynamics; the states'
lack of clear information when negotiating and concluding the agreement; and
situations where the design is drawn from an absolutely different context.

4 Middle East: the kingdom of political economy

In particular, four broad claims have been made about the applicability of International Relations to this region: that the region has to be seen in terms of the pattern of its historical incorporation into the global political and economic system, 'differential integration', and that it is this which defines the character, and limited powers, of regional states; that the central category for understanding the international relations of the Middle East and its relations with outside powers is the institutional, rather than juridical concept, of the state, inviting, but leaving open a study of the influence on its decision-making processes and policy-making; that the international politics of the region have to be seen at three distinct levels, in terms of the interaction of global structures of power, of regional states and of non-state or social movements; and that the belief systems, ideologies, norms of the region while they draw selectively on the past are not traditional but modern phenomena that have to be related to the interests of these contemporary states and their apparatus.

Fred Halliday, The Middle East in International Relations

Seven bilateral treaties in the Middle East dealing with sea-related subjects are relevant to this work: the boundary agreement between Bahrain and Saudi Arabia of 22 February 1958; the agreement on the delimitation of the offshore and land boundaries between the Kingdom of Saudi Arabia and Qatar of 4 December 1965; the agreement concerning the sovereignty over the islands of Al-'Arabiyah and Farsi and the delimitation of the boundary separating submarine areas between the Kingdom of Saudi Arabia and Iran of 24 October 1968; the agreement between Qatar and Abu-Dhabi on the settlement of maritime boundaries and the ownerships of islands of 20 March 1969; the agreement concerning the delimitation of the continental shelf between Iran and Bahrain of 17 June 1971; the agreement concerning the delimitation of the continental shelf between Iran and Oman of 25 July 1974; and the agreement on the delimitation of boundaries between Saudi Arabia and the United Arab Emirates of 21 August 1974.[115]

The last treaty analysed here was signed between Saudi Arabia and the United Arab Emirates in 1974. There had been a positive trend in negotia-

---

115 Available at http://www.un.org/Depts/los/LEGISLATIONANDTREATIES/persi an_gulf.htm.

tions and legally binding documents over the preceding sixteen years. Somehow, it had been possible to foresee solutions to territorial sovereignty issues and from there to encourage joint actions. Nevertheless, actions were subsequently delayed until in 1979, everything between the states had been reduced to mere text provisions. At that point, the question seems to be whether cooperation was overwhelmed by regional warfare and international political economy, or whether the treaties had been dead letters from the beginning.

Institutional designs

Each treaty analysed here aimed was at forging a consolidation of sovereignty (Table 2.1 provides an overview of matching design elements). In this sense, the outcomes of negotiations had to demonstrate equality in the states' authority, *tête à tête*, in order to preserve the respective governments' national prestige and legitimacy. In addition, treaties had to open the possibility of solving historical conflicts and open the chances of future material gains.

	Saudi Arabia and Bahrain	Saudi Arabia and Qatar	Saudi Arabia and Iran	Saudi Arabia and UAE	Qatar and Abu Dhabi	Iran and Bahrain	Iran and Oman
NORMS	Liberal Economic	Liberal Economic	Liberal Economic	Liberal Economic	Liberal Economic	Neutral Liberal Economic	Neutral Liberal Economic
SCOPE	Narrow; unintrusive	Narrow; Unintrusive	Narrow; unintrusive; additional goal	Narrow; unintrusive	Narrow; unintrusive	Narrow; unintrusive	Narrow; unintrusive
MANDATE	Outcome oriented; distributive;	Outcome oriented; distributive;	Outcome oriented; distributive;	Outcome oriented; distributive;	Outcome oriented; distributive;	Outcome oriented; distributive;	Outcome oriented; distributive;
FORMAL RULES	Programmatic	Symmetrical; programmatic	Symmetrical; programmatic	Symmetrical; programmatic	Symmetrical; programmatic	Symmetrical; programmatic	Symmetrical; programmatic

Table 2.1

Norms

The treaties under analysis are drafted in secular terms in most of their preambles and all of their main texts.[116] In this sense, they express a preliminary 'spirit of affection and mutual friendship',[117] or 'links of friendship and bonds of brotherhood',[118] but confine themselves to legal terms such as: 'due respect to the principles of the law and particular circumstances',[119] and 'just, equitable and precise' terms and conditions.[120] Even the treaty between Saudi Arabia and the United Arab Emirates – which has a strong pro-Arab identity introduction stating that it was signed '[i]n pursuance of the principles of the Holy Shariah professed by the Islamic Community, proceeding from the spirit of Islamic solidarity that embraces the Arabian Peninsula, and on the basis of the bonds of amity between them, the links of brotherhood between their fraternal peoples and the relationship of neighborliness existing between their two countries',[121] does not venture into religious meanings or terms in the rest of the text, and focuses on precise definitions and the implementation of the principles of sovereignty and equity, which are usually accompanied by the precise geographic locations of boundary limits, divisions of sovereignty rights, hydrocarbon property rights or development conditions, or reference to technical commissions, as necessary. These elements are part of Economic Liberalism as well as reflecting of pro-sovereignty tendencies (*see* Table 2.2).

Scope

Every treaty treats sovereignty issues: boundary and territorial delimitation and sovereignty rights over natural resources. Additionally, the treaty between Saudi Arabia and Iran has a *symbolic* effect – if I may – in 'for-

---

116 The two exceptions to neutral forms in the preliminary lines are the treaty between Saudi Arabia and the UAE.
117 Treaty between Saudi Arabia and Bahrain (1958).
118 Treaty between Saudi Arabia and Qatar (1965).
119 Treaty between Saudi Arabia and Iran (1968).
120 Treaty between Iran and Bahrain (1971).
121 Treaty between Saudi Arabia and United Arab Emirates, preamble, paragraph 2.

mally' putting an end to a long-lasting difference regarding sovereignty over the islands of Al-Arabishah and Farsi.[122]

Formal Rules

The treaties are programmatic and as such, flexible as to the state's eventual needs and political will. To begin with, all of them make precise definitions of boundary coordinates; some describe sovereignty areas or 'prohibited areas' for oil development. However, ultimately, all these technical issues have to be confirmed by third parties (companies, third state officials or joint commissions) upon consensual initiative of the states parties. There is a second form of programmatic substance, when the agreement declares options for oil development 'in the event' that an exploitable natural resource is found. Once again, this action is to be agreed upon consensus (*see* Table 2.2).

Mandate

The mandates of the treaties are product-oriented and distributive. They intend to define territorial sovereignty, determining sea and land boundaries, continental shelf limits, the ownership of islands, and natural resource (oil) development rights.

---

122 *Id.*

	Saudi Arabia and Bahrain	Saudi Arabia and Qatar	Saudi Arabia and Iran	Saudi Arabia and UAE	Qatar and Abu Dhabi	Iran and Bahrain	Iran and Oman
COMMISSIONING TO DEFINE BORDERS	No	Yes (Art. 3 and Art. 5)	Yes (Art. 3.b)	Yes (Art. 5.3, Art. 6, and Art. 7)	No	No	Yes (Art. 3)
PROHIBITION AREAS FOR OIL DEVELOPMENT	No	No	Yes (Art. 4)	No	No	No	No
OIL DEVELOPMENT SYSTEM	Unitisation (Art. 2.7)	None	Geological Cooperation (Exchanged letter I.b and II.b)	None	Joint Operation (Art. 6.)	Joint Operation OR Unitisation (Art. 2)	Joint Operation OR Unitisation (Art. 2

Table 2.2

The factors influencing institutional designs

Agreeing with Halliday, I find that the seven institutions analysed here – designed between 1958 and 1974 – were forged in a 'kingdom of international political economy'.[123] In effect, the states of Middle East seem to have developed a faculty to divorce what can be understood as 'pure' politics from 'political economy', which has had an impact from the end of the Second World War to the present.

From the late 1940s, the Cold War was prosecuted within a global, strategic, ideological and economic context, and was determined, as far as the Middle East was concerned, from the outside. The impact of the Cold War was, to a considerable extent, shaped by the states and societies which were already established in the region. It had, therefore, a profound effect on the Middle East, its states and peoples, and on the place of the Middle East within the international system as a whole. However, it is important to distinguish here between the Cold War as a global, formative context, and the Cold War as a system of strategic control which dictated the actions of local states and movements. As states and political movements manoeuvred to take advantage of the global rivalry, that rivalry itself had a profound impact on many parts of the region, inspiring mass movements of the left and right. External forces sought allies and poured weapons, advice and in some cases economic assistance into the region. In addition, the Cold War certainly accelerated the transition in international involvement that the Second World War had begun – pushing out the French and the British, bringing in the Americans and the Russians. Yet neither East nor West ever found it easy to influence their allies, Arab, Israeli, Turkish or Iranian.

---

123 'Political economy asserts an indissoluble interconnection of political factors – states, conflict, ideology – with the economic – production, finance, technology. It is, indeed, only since the late nineteenth century that the conceptual, and academic, separation between them has been enforced. In every region of the world economic issues, domestic and international, are inseparable from politics. Yet this is perhaps nowhere more true then in the Middle East. In republics and monarchies alike politics and political aspiration, not to say fantasy, are the key, inter alia, to the economic projects, supposedly "planned", "constructed" and the like, which are promulgated by sometimes megalomaniac rulers through their industrialisation and other programmes'. FRED HALLIDAY, THE MIDDLE EAST IN INTERNATIONAL RELATIONS: POWER, POLITICS AND IDEOLOGY, 261 (2005).

Much of what took place in the inter-state relations of the Middle East during the Cold War had, moreover, little to do with the global conflict. Apart from the inter-Yemeni rivalry, the major inter-state conflicts – Arab–Israeli, Iran–Iraq (from 1975), Syrian–Lebanese – had only an indirect relationship in their origins and outcomes to the Cold War. If this independence of regional actors was true for states, it was even more so for the social and political forces which states did not control. A final verdict on the past may rest not with global strategy, but with the international political economy. What was decisive for most people in the region – the pursuit of a livelihood and a measure of economic security – had almost nothing to do with the Cold War: the USSR never offered significant amounts of investment or aid, let alone a different economic model, whilst the monies coming from the West were largely channelled to elites through the provision of oil revenues from consumers and then recycled back to London and New York. Thus the dynamic of Middle Eastern events, while certainly autonomous of the world conflict that was the Cold War, was also loosely interconnected with it.

Political autonomy can partially explain two important facts about the six parties to the offshore international agreements under study (Saudi Arabia, Kuwait, Qatar, Bahrain, Oman, and Iran): first, none took direct part in the region's major conflicts which involved non-regional states (i.e., the Suez Crisis, the 1967 Six-Days War, and the 1973 Yom Kippur War); and second, excepting Iranian unilateral initiatives, there were no armed hostilities between them.[124] This may have been a stimulus for the

---

[124] According to Fred Halliway, '[t]he British withdrawal from the Gulf, begun in Kuwait in 1961 and effected in the other smaller lower Gulf states a decade later, led the Shah to project Iran as the new dominant power in that region. Social upheaval to west and east, therefore, as much as the strategic maneuvering of the Cold War, altered Iran's international perspective. By the 1970s Iran's strategic orientation was southwards, not to the north. In the 1970s, in line with this vision, the Shah took a number of military initiatives: from 1969 to 1975 Iran fought a low-level but persistent border war with Iraq, in effect the first "Gulf war". This war was only ended with the Algiers Agreement of 1975 in which the land and water frontiers of the two states were settled, and a pledge of mutual non-interference provided. This confrontation presaged interventionist action elsewhere. In 1971, on the eve of the British withdrawal, Iranian forces occupied three Arab islands, Abu Musa and the greater and Lesser Tunbs, belonging to the United Arab Emirates. In 1973 Iran sent several thousand counter-insurgency troops to the southern Omani province of Dhofar to fight revolutionary guerrillas active there. To the south-east it provided support to counter-insurgency campaigns by

seven arrangements, and a hint at why they are enlightened by such strong economic pragmatism.

Most theories and tendencies have supporters and opponents, and the final position exposed here is not an exception. In my case, I am of the belief that the following can be considered as the 'most influential' factors affecting the agreements' negotiation and design: (i) national politics (political semi-authoritarian regimes and economic nationalism); (ii) the evolution of the oil market; (iii) the evolution of international law; and (iii) the evolution of the oil industry. All these variables experienced a strong and consolidated change in the period 1958 to 1974.[125]

Local politics: formation and characteristics

The choice taken here is to understand the international relations of the Middle East and its relations with outside powers as an expression of the 'institutional' rather than 'juridical' concept of the state, which leaves open the study of its influence on decision-making processes and policy-making.[126] In this sense, the relations between states, external and regional, in the 1918–1939 period may have been less important for later history than what was occurred within the states. It was not in the Great Power politics nor in the growing inter-state relations in the region, but in the less visible but potentially equally significant domain of the formation of states and societies that the post-1918 period was to prove so important for later events. In large measure, the political, social and international formation of the region took place in this period. It was through this internal process of change that the social, ideological and hence political dimen-

---

Pakistan against rebels in Baluchistan', *supra* note 9, at 103. *Also*, Mohamed Abdullah Al Roken, *Dimensions of the UAE–Iran Dispute over Three Islands*, in UNITED ARAB EMIRATES: A NEW PERSPECTIVE (IBRAHIM AL ABED and PETER HELLYER eds., 2007.).

125 Unintentionally, the analysis offered in this chapter fits historical sociology theory, an approach applied in particular to the Middle East by Fred Halliday, *supra* note 9, at 287. To read further down this optional path, *see* ANTHONY GIDDENS, THE NATIONAL STATE AND VIOLENCE (1985); *also*, CHARLES TILLY, THE FORMATION OF NATIONAL STATES IN EUROPE (1975).

126 *See* Halliday, *supra* note 9, at 200.

sions of the modern Middle East were so decisively shaped.[127] Of course, the recurrence of external, Great Power military involvement in the region and of war between regional states was accompanied by irregular war within states, the actions of opposition forces contesting external involvement and that of local states alike.[128] Once this concept is clearly understood, four processes merit particular attention. One is the creation of modern state institutions. Defining the map of the Middle East in the years immediately after 1918 provided a set of empty boxes or shells within which these states could, and did, develop as institutions of power and appropriation, with aspirations both internal and external. These were institutions of administration and coercion, run by colonial and then nationalist powers, which imposed increasingly effective control over the territories they came to rule. They provided employment for growing numbers of people and came to direct society – economic development and education – according to the wishes of their rulers. Of no little importance for the later politics of these countries, the inter-war years saw the development of armed forces: these were the first institutions in these societies to be attuned to modern values and, with their now distinct social and economic interests, increasingly became aspirants to political power and to the status of defenders of the nation.

Secondly, the heads of states embarked, as part of their attempt to assert control and to forge more effective and malleable political communities,

---

127 This pattern of autonomous inter-state war had already been set in the inter-war period, when the rising state of Saudi Arabia seized power by conquering much of the Arabian Peninsula by the sword and then, in 1934, fought a war with the only other independent Arab state at that time, Yemen. The Saudis prevailed with the Treaty of Taif (1934), but in imposing a peace on Yemen that forced the latter to cede three provinces hitherto considered as part of their national territory (Asir, Jizan and Najran), they laid the basis for a Saudi–Yemen conflict, and much Yemeni resentment, which was to last for decades. Only in June 2000, at a summit between Prince Abdullah and President Al Abdullah Salih, was a comprehensive agreement on this frontier, the longest undelineated one in the world, reached.

128 Such actions ranged from substantial guerrilla campaigns involving the mobilisation of thousands of armed men and the occupation of territory, as in Algeria (1954–1962), Turkey (1984–1998), Iraq (intermittently 1958–2003) and Yemen (1962–1970), to war between confessional groups within one country (Lebanon, 1979–1990), to lower-level campaigns of bombing and assassination (Egypt, 1990s; Iran from 1981 onwards), through to more sporadic armed actions linked to political campaigns (the Second Palestinian Intifada, 2000 onwards).

on the construction of national identities as a basis of statehood.[129] Part of this involved the assertion and maintenance of claims with regard to other states, based on what were viewed as historic rights, or on denunciation of the partitions and divisions imposed by colonialism (Egypt claimed Sudan, Syria, Lebanon, Iraq, Kuwait, Saudi Arabia the rest of the Peninsula). In every case, the states were delineated but the boundaries unfinished.

Thirdly, these states embarked, or thought they had embarked, on a process of cultural and ideological change, closely linked to the consolidation of their own power: this state-directed change in society promoted a certain form of secularisation. Secularisation reflected a commitment to the values of the modern world, as exemplified in Europe; but in the modernisation of Middle Eastern states, secularism was not part of a process of building tolerance between communities or of creating a civic and legal space independent of the state, as it had been in areas of Europe. Secularisation was, above all, a policy intended to strengthen states: it was a reflection of the desire of these states to reduce or break the power of an alternative centre of power exercised by influence on education, land and law, and to forge a new ideology of control over society.

All of these three processes – state formation, nationalism and secularisation – were changes brought about from above and in response to external pressures: of equal importance was a fourth process occurring below. The final years of the Ottoman Empire and even more so, the years following the imposition of the post-1918 settlement, saw the emergence in a range of countries of popular movements, combining social with economic demands.[130] These created a context in which both colonial rulers and incumbent states faced challenges from below, to which they replied with

---

129 Education was a central means for the promotion of these new identities and rested upon the creation of a national history, drawing where available on both Islamic and pre-Islamic elements. While each state sought to assert its own individual identity and historic validity, each also made a claim to be part of wider communities.

130 Prior to the First World War, there had been major upheavals in Iran (the Constitutional Revolution of 1906–1908), Turkey (the Young Turk revolt of 1908 and its aftermath) and Egypt (1907). Armenian nationalism and those of the Balkans had also become more assertive and violent. In the years after 1918 there were local and nationalist uprisings in Egypt (1919), Iraq (1920), Syria (1925) and Palestine (1936), as well as in Morocco (1926). The Kurds of Turkey, Iraq and Iran also rose in revolt against military ruler and colonial power alike.

a combination of coercion and cooptation. These revolts contested external domination and, in the case of Palestine, immigration. They also challenged incumbent social and political elites, some only just installed, who were viewed as being linked to the colonial powers. Inevitably, the processes would produce a counter-reaction, one which was decades later to emerge in the form of an Islamist, or fundamentalist, politics which challenged the power of secular states throughout the region.[131]

Little by little, the political elites took advantage of these four processes, ending up controlling all land and oil revenues, which in turn played a decisive role in the maintenance of their leadership after the oil market took off. To be precise, the core characteristic of oil management in the Middle East was that its revenues were paid to the state, meaning that it was those who controlled the state – the ruling families in the Gulf and the military elites – who disposed of the money. A combination of an established 'clientilism' and a refusal of the authoritarian regimes to submit their accounts to public scrutiny led to a situation in which a considerable proportion of oil revenues were, as euphemistically expressed by the US embassy in Saudi Arabia, 'off budget'. A rough rule of thumb applicable to republics (Iraq and Yemen) and to monarchies alike, was that a third of the income from oil, and a larger percentage of the revenue from foreign investments, went to the ruling family.

As we can see, by the end of the 1950s, Saudi Arabia, Iran, Kuwait, Oman, Qatar, Bahrain, and Abu Dhabi (UAE) had a consolidated internal regime and a firm position on the map. Their next step was to explore how to increase their material benefits.

The evolution of the oil market: international and regional interests

Several consequences across the entirety of the international political economy and the Middle East have followed from the evolution of the oil market from its beginnings. Firstly, oil shaped the social character of the state and, by derivation, of the economy. The states which possessed oil derived from it an increasingly important income and came to depend largely on that income; they were what were termed 'rentier' or more respectfully, 'distributive' states. While these revenues greatly strength-

---

131 *See* Halliday, *supra* note 9, at 229.

ened these states and enabled them to increase their imports, this reliance had, as we shall see, other, negative consequences, ones that any economic historian or social scientist writing about any state or society in the past five hundred years could have anticipated. In all cases where unearned income has come to dominate state revenues, a pattern marked by parasitism and factionalism has emerged.

Secondly, while it was not in itself a source of conflict, oil did provoke a political reaction within these societies: the development of the oil industry and the issue of ownership of oil became objects of great dispute within Middle Eastern countries, as well as between these countries and the West. As Western policy in the Middle East was increasingly regarded as being dictated by the desire to control oil production, this led to strong nationalist reaction.

Between 1950 and 1973, the oil industry outside the Middle East grew ninefold – a rate of increase of ten percent per year sustained, amazingly, over a period of twenty-three years. Two hundred new refineries were built outside the United States and some older ones expanded. About 1,750 tankers of ever-increasing size were launched. Over 2.5 billion new motor vehicles were put on the roads, more than half of them in the United States. Air travel replaced ocean liners and every other form of long-distance travel. In Western Europe and Japan oil dethroned coal. Petrochemicals brought a wide range of entirely new products. World oil demand including the United States and the Centrally Planned Economies (CPEs) more than quintupled, growing from 11 million barrels per day (mbd) in 1950 to 57 mbd in 1970. Economies boomed, peace (more or less) prevailed. In the oil industry, the seven major oil companies (Chevron, Exxon, Gulf, Mobil, Texaco, British Petroleum, and Royal Dutch/Shell), the 'Seven Sisters', reigned supreme. It was the golden age of oil. People's lives were transformed and the transformation was possible for two reasons: first, the basic technology already existed[132] and second, after the 'Big Bang', the oilfields of the Middle East provided the resource-base for low-cost petroleum products across the board – fuels, lubricants, asphalt and petrochemical feedstocks.

The oil industry's 'Big Bang' was the discovery and development of the super-giant fields of the Middle East. Enough of them had been discov-

---

132 The internal combustion and diesel engines had been invented in the nineteenth century, the jet in the 1930s, the basic refining processes were well-established, thermal cracking had been introduced in 1913, and catalytic cracking in 1936.

ered well before 1950, where we take up the story, to bring some countries and companies to the realisation that they were on the brink of a new world of oil of some kind. In Iran the Masjid-e-Sulairnan field had been discovered as early as 1908, to be followed much later by Gach Saran (1928) and Agha Jari (1938), all giants. Though not fully proved until later, BP (AIOC at the time) had in fact more reserves in these three fields than the combined reserves of the entire United States. In north-central Iraq, the great Kirkuk field was discovered in 1927 and a twelve-inch pipeline to Haifa in Palestine and Tripoli in Lebanon was completed by 1935, looped with a sixteen-inch line in 1950. Ain Zalah was discovered in 1939 and Zubair in 1948. A thirty to thirty-two-inch line from Kirkuk to Banias in Syria was completed in 1952, and a twelve-inch line from Zubair to Fao on the Persian Gulf in 1950. Super-giant Burgan was discovered in Kuwait in 1938. Despite the total lack of infrastructure where the fields were discovered across the Middle East, all were cheap to produce because they were so prolific, but Kuwait was the cheapest of all. Near the coast but at some elevation from it, oil flowed under great reservoir pressure to the surface, then downhill through the pipelines to the shipping terminal. By 1960, with operating expenses of less than 5 cents per barrel, Kuwait oil sold for USD 1.50/B. In Saudi Arabia, the Damrnam field had been discovered in 1938, followed by Abqaiq and Abu Hadnya in 1940, Qatif in 1945 and Ghawar, the world's largest oilfield, in 1948. Other, smaller fields had also been discovered in Bahrain (Awali in 1932) and Qatar (Dukhan in 1940).[133]

The bountiful supplies of the Middle East were of course not enough in themselves; the oil still had to be refined and brought to market at capital costs, by comparison with which the investment required for the development of the Middle East was nugatory. Throughout the period from 1950 to 1970, oil markets grew rapidly though unevenly from country to country and area to area, as might be expected. Overall, as noted above, they more than quadrupled. The US market, already relatively well developed, more than doubled, Western Europe and the CPEs both grew ninefold; Japan, one hundredfold, increasing from virtually no consumption at all to become, with a demand of 4 mbd, the world's second largest consumer of oil products after the United States. The rest of the world – that is, the

---

133 *See* FRANCISCO PARRA, OIL POLITICS: A MODERN HISTORY OF PETROLEUM 33 (2004).

developing countries – increased their demand fourfold. By the mid 1960s, such was the growth in demand that companies were having to reassure the public that there would be enough oil in the long term to supply the rising tide of consumption.[134]

---

134 *Id.*, at 41.

MAJOR OIL EXPORTERS (in millions of barrels)

Year	Kuwait	Saudi Arabia	Iran	Iraq	Abu Dhabi	Qatar	Others	Total Middle East	Libya	Total ME & Libya	Algeria	Nigeria	Venezuela
1955	400.8	850.4	110.7	289.6		40.8	11.0	1,152.8		1,152.8			788.6
1960	608.0	473.8	856.2	339.0		62.5	16.4	1,855.9		1,855.9			982.7
1961	624.4	530.2	397.5	347.2		64.2	16.4	1,979.9	5.1	1,985.0			1,009.0
1962	703.7	589.3	448.0	347.6	5.5	67.8	16.4	2,178.3	59.5	2,237.8			1,101.9
1963	749.7	688.2	499.8	402.8	17.6	70.7	16.4	2,395.2	167.2	2,562.4			1,121.9
1964	852.2	684.5	580.4	440.6	68.0	77.6	17.9	2,721.2	313.7	3,034.9			1,175.8
1965	850.6	787.4	644.1	458.8	102.3	83.3	20.3	2,946.8	442.6	3,389.4			1,187.2
1966	902.5	981.6	729.0	485.0	132.5	105.5	21.9	3,308.0	547.4	3,855.4			1,161.8
1967	905.4	1,004.7	898.1	424.0	137.7	116.8	45.1	3,526.8	621.1	4,147.9	289.9		1,226.9
1968	951.6	1,039.3	975.9	525.2	181.4	124.8	114.8	3,972.5	945.0	4,917.5	330.5		1,235.3
1969	1,005.7	1,157.7	1,158.5	529.2	218.8	125.4	151.7	4,347.0	1,131.5	5,478.5	358.5		1,244.9
1970	1,081.5	1,359.4	1,817.7	544.9	253.5	133.4	179.8	4,870.2	1,187.6	6,057.8	276.0	876.0	1,288.0
1971	1,168.7	1,707.8	1,562.2	593.5	338.5	156.5	179.0	5,705.7	988.9	6,694.6	373.0	531.4	1,205.9
1972	1,175.7	2,162.7	1,751.9	381.6	384.5	176.8	184.6	6,217.0	812.9	7,029.9		628.0	1,132.8
Total 1963-1972	**9,648.6**	**11,532.8**	**10,112.6**	**4,785.6**	**1,834.5**	**1,169.8**	**931.5**	**40,010.4**	**7,157.9**	**47,168.3**			**11,941.2**

Source: Petroleum Information Foundation (New York), Background Information, Paper No. 16, October 1973.

Table 2.3

Year	Kuwait	Saudi Arabia	Iran	Iraq	Abu Dhabi	Qatar	Others	Total Middle East	Libya	Total ME & Libya	Algeria	Nigeria	Venezuela
1955	307.7	287.8	90.5	206.5		34.1	9.0	934.9		934.9			596
1960	465.2	355.2	285.3	266.3		54.0	13.0	1,439.0		1,439.0			877
1961	464.3	400.2	301.2	265.5		53.3	13.0	1,497.5	3.2	1,500.7			938
1962	526.3	451.1	333.8	266.6	2.8	55.8	13.0	1,649.4	38.5	1,687.9			1,071
1963	556.7	502.1	398.1	325.1	6.4	59.5	13.0	1,860.9	108.8	1,969.7			1,106
1964	655.0	561.0	469.7	353.1	12.4	65.5	14.3	2,131.0	197.4	2,328.4			1,122
1965	671.1	655.2	522.4	374.9	33.2	68.5	16.4	2,341.7	371.0	2,712.7			1,135
1966	707.2	776.9	593.4	394.2	99.8	92.1	18.5	2,682.1	476.0	3,158.1			1,112
1967	717.6	852.1	736.7	361.2	105.0	101.8	23.6	2,898.0	631.0	3,529.0	191.1		1,254
1968	765.6	965.5	817.1	476.2	153.2	109.5	83.1	3,370.2	952.0	4,322.2	261.8		1,253
1969	812.2	1,008.0	937.8	483.5	191.1	115.2	118.2	3,666.0	1,132.0	4,798.0	298.8		1,289
1970	895.1	1,199.7	1,136.3	521.2	233.1	122.0	150.2	4,257.6	1,294.8	5,552.4	325.0	411.0	1,406
1971	1,399.8	2,148.9	1,944.2	840.0	430.7	197.8	19.6	7,154.0	1,766.0	8,920.0	350.0	915.0	1,702
1972	1,656.8	3,106.9	2,379.8	575.0	550.9	254.8	222.6	8,746.8	1,598.0	10,344.8	700.0	1,174.4	1,948
Total 1963-1972	8,837.1	11,776.3	9,935.5	5,049.4	1,815.8	1,186.7	852.5	39,108.3	8,527.0	47,635.3			13,327

Source: Petroleum Information Foundation (New York), Background Information, Paper No. 16, October 1973.

Table 2.4

Consumption, demand and revenues increased (see Tables 2.3 and 2.4),[135] but only for the major companies. The position was extremely disadvantageous for oil-producing countries, as none had the skills, technology or marketing networks necessary to exploit their oil riches. Therefore, the contracts negotiated between the major oil companies and their governments granted particular oil firms or groups of firms exclusive rights to explore for oil in an agreed territorial area within each country.[136] Revenues thus remained uneven until some states challenged these traditional arrangements. Flexing the sinews of their newly-won political independence, achieved after many years of struggle, the Arab oil-producers sought changes in the contracts at the expense of the oil companies,[137] and then moved a step further – as will be detailed later in this chapter – nationalising oil companies in various forms.[138] The last measure was taken in 1960, when at the initiative of Venezuela, four other major oil-producing countries, Saudi Arabia, Kuwait, Iran and Iraq, created OPEC with the primary objective of improving their bargaining position *vis-à-vis* the major oil companies.

OPEC was founded with the express purpose of coordinating the petroleum policies of its members and therefore safeguarding their individual and collective interests. It was a direct response to the anger and frustration felt among Middle Eastern oil-producers over the major oil companies' decision to lower the posted price of oil in the face of a global glut of petroleum. With their oil revenues seriously threatened, the Arab members of OPEC denounced the oil companies' decision to reduce the price of oil and pushed for greater control over production and pricing policies. To this end, they sought a more direct and active role in the exploita-

---

135 Taken from Zuhayr Mikdashi, *Cooperation among Oil Exporting Countries with Special Reference to Arab Countries*: A Political Economy Analysis, 28 and 32 International Organization, 1.

136 *See* VALÉRIE MARCEL, OIL TITANS, 13 (2006),

137 Changes included tax and shares. *See* VALÉRIE MARCEL, *supra* note 13, at 19. *Also* Parra, *supra* note 19, at 13. These measures were opposed not only by the big oil companies themselves, but also by Western governments, who saw the oil firms as agents of their national interests and the key guarantors of the Global North's access to oil, a critical energy resource. Examples of this opposition can be seen in events such as the assassination in 1953 of Muhammad Mosadeq, leader of the movement to nationalise oil in Iran by agents of the United Kingdom and the United States intelligence services. *See* Parra, *supra* note 19, at 25.

138 *See* VALÉRIE MARCEL, *supra* note 13, at 25.

tion and marketing of their oil resources. Each country realised, however, that collaborative efforts were required to bring fundamental change to the oil regimes, and that the oil production and pricing policies among OPEC members needed to be coordinated and unified to achieve the benefits the organisation was desired to achieve generally, and for each member-state individually.[139]

Oil certainly transformed the position of the Middle East in the world economy and whatever else it may have done, it guaranteed continued attention by the outside world to the affairs of the region. In particular, as the United States – which had previously been largely self-sufficient in oil – became a major energy importer from the 1970s, so the power of the Middle Eastern producers in the world market became stronger.

The evolution of International Law: sovereignty over natural resources

The nationalisation by Iran of the Anglo-Iranian Oil Company in 1951 marked the first major economic North–South clash in the post-war period. By the terms of a concession agreement concluded in 1933 between the British-owned Anglo-Persian Oil Company and the government of Iran, the company had acquired the exclusive right to extract and process petroleum in a specified area in Iran until 1993. On 1 May 1951, Dr. Mossadegh, the Prime Minister of the then socialist Iranian government, announced the official decision to nationalise the company and to annul the 1933 oil concession agreement. The National Iranian Oil Company was established to take over the exploitation of the nationalised oilfields. Obviously, this situation jeopardised the free flow of oil to the United Kingdom. When the Iranian government announced its unwillingness to submit the dispute arising from this act of nationalisation to arbitration, on 27 May 1951 the British government filed an application with the International Court of Justice. It requested the court to declare that the Iranian government was under an obligation to submit the dispute to arbitration under the provisions of the arbitration agreement or, alternatively, that the Iranian government was acting contrary to international law, particularly to its obligations under the 1933 Agreement. In a provisional

---

139 See Euclid A. Rose, *OPEC's Dominance of the Global Oil Market: The Rise of the World's Dependency on Oil*, 58 MIDDLE EAST JOURNAL 424. *Also* Parra, *supra* note 19, at 89.

order of 5 July 1951 the court prescribed a number of interim measures for the parties to observe to prevent aggravation of the dispute, including permission to continue the operations of the Anglo-Iranian Oil Company pending settlement of the dispute. Iran, however, refused to comply with these interim measures. Accordingly, in October 1951 the British government then requested the Security Council to order Iran to obey the provisional court order (cf. Article 94.2 of the UN Charter). After an extensive debate the Security Council decided, however, following a French proposal, to adjourn its debate on the issue until the court had ruled regarding its jurisdiction on the matter. On 22 July 1952, the court passed final judgment, by nine votes to five, concluding that it had no jurisdiction to deal with the case. One of its main grounds was that the concession contract did not fall within the meaning of the term 'international conventions' of Article 38(1) of the court's Statute. The court stated that the agreement itself could not be considered as anything more than a 'concessionary contract between a government and a foreign corporation'. In 1953 a new Iranian government under the leadership of the Shah came to power after a *coup d'etat* in which the British and US secret services were reportedly involved. In 1954 the new government signed a new agreement with an international oil consortium consisting of British, Dutch, American and French oil companies.

During the years of the Anglo-Iranian dispute, ECOSOC and the UN General Assembly discussed the right of peoples and nations to take charge of their own natural resources. This occurred in two different contexts, namely the debates on: (a) the promotion and financing of economic development in under-developed countries; and (b) the drafting of human rights treaties. The first context has often been overlooked, both in the international law literature and in relevant UN documents.

From the 1960s, developing countries actively pursued the implementation of the principle of permanent sovereignty over natural resources because they perceived this to be a key driver of their economic development and of the redistribution of wealth and power in their relations with the industrialised world. Consequently, during the period 1963–1970, emphasis on State control and the actual ways of implementing the principle of permanent sovereignty over natural resources attracted increasing attention, and the link between permanent sovereignty over natural resources and promoting the development of host countries was firmly established. General Assembly Resolution 2158 (XXI), in particular, was instrumental in this, both substantively and politically. The guidelines it provides for relations and cooperation between foreign investors and developing host countries contain a relevant and substantive inventory of the problems involved and recommend constructive policies. Politically, it was important that the broad coalition which supported the 1962 Declaration on Permanent Sovereignty Over Natural Resources persisted over time in the subsequent documents (see the Detail of Resolutions, below).

Nevertheless, after 1970 the debate on permanent sovereignty over natural resources changed substantially. Some controversial elements were more vigorously introduced in the discussions on the scope and content of permanent sovereignty, in particular the creeping jurisdiction of coastal States over adjacent sea areas and marine resources, which often led to fishery disputes and allegations of economic coercion. Furthermore, in the early 1970s an increasing number of developing countries nationalised certain sectors of their economies, including foreign-owned sectors, and sought international legitimisation for these actions through the United Nations, leading to a resumption of the nationalisation debate.

DETAIL OF RESOLUTIONS

General Assembly resolutions on permanent sovereignty over natural resources

GA Resolution	Date of adoption	Voting record	Title
523 (VI)	12 January 1952	Adopted unanimously	Integrated Economic Development and Commercial Agreements
626 (VII)	21 December 1952	36 (60%) – 4 – 20	Right to Exploit Freely Natural Wealth and Resources
837 (IX)	14 December 1954	41 (75%)–11 – 3	Recommendations Concerning International Respect for the Right of Peoples and Nations to Self-Determination
1314 (XIII)	12 December 1958	52 (69%)–15 – 8	Recommendations Concerning International Respect for the Right of Peoples and Nations To Self-Determination
1720 (XVI)	19 December 1961	85 (94%) – 0 – 5	Permanent Sovereignty over Natural Resources
1803 (XVII)	14 December 1962	87 (86%) – 2–12	Permanent Sovereignty over Natural Resources
2158 (XXI)	25 November 1966	104 (95%) – 0 – 6	Permanent Sovereignty over Natural Resources

GA Resolution	Date of adoption	Voting record	Title
2386 (XXIII)	19 November 1968	94 (91%) – 0 – 9	Permanent Sovereignty over Natural Resources
2692 (XXV)	11 December 1970	100 (92%) – 6 – 3	Permanent Sovereignty over Natural Resources of Developing Countries and Expansion of Domestic Sources of Accumulation for Economic Development
3016 (XXVII)	18 December 1972	102 (82%) – 0 – 22	Permanent Sovereignty over Natural Resources of Developing Countries
3171 (XXVIII)	17 December 1973	108 (86%)–1–16	Permanent Sovereignty over Natural Resources
3201 (S-VI)	1 May 1974	Adopted without vote	Declaration on the Establishment of a New International Economic Order
3202 (S-VI)	1 May 1974	Adopted without vote	Programme of Action on the Establishment of a New International Economic Order
3281 (XXIX)	12 December 1974	120 (88%) – 6–10	Charter of Economic Rights and Duties of States
32/176	19 December 1977	130 (94%) – 0 – 8	Multilateral Development Assistance for the Exploration of Natural Resources
33/194	29 January 1979	Adopted without vote	Multilateral Assistance for the Exploration of Natural Resources

General Assembly resolutions relevant to the question of sovereignty over natural resources

GA Resolution	Date of adoption	Voting record	Title
1514 (XV)	14 December 1960	89 (91%) – 0 – 9	Declaration on the Granting of Independence to Colonial Countries and Peoples
1515 (XV)	15 December 1952	Adopted unanimously	Concerted Action for Economic Development of Economically Less Developed Countries
1813 (XVII)	18 December 1962)	Adopted unanimously	Economic Development and the Conservation of Nature
2626 (XXV)	24 October 1970	Adopted without vote	International Development Strategy for the Second United Nations Development Decade
2849 (XXVI)	20 December 1971	85 (70%) – 2 – 34	Environment and Development
2995 (XXVII)	15 December 1972	115 ((92%) – 0–10	Co-operation between States in the Field of Environment

GA Resolution	Date of adoption	Voting record	Title
3129 (XXVIII)	13 December 1973	77 (62%) – 5 – 43	Co-operation in the Field of Environment Concerning Natural Resources Shared by Two or More States
3362 (S-VII)	16 September 1975	Adopted unanimously	Development and International Economic Co-operation
3517 (XXX)	15 December 1975	123 (94%) – 0 – 8	Midterm Review and Appraisal of Progress in the Implementation of the International Development Strategy for the Second United Nations Development Decade
34/99	11 December 1979	Adopted without vote	Development and Strengthening of Good Neighborliness Between States
34/186	18 December 1979	Adopted without vote	Co-operation in the Field of Environment Concerning Natural Resources Shared By two or More States
35/7	5 December 1980	Adopted without vote	Question of the Draft World Charter of Nature
35/56	28 October 1982	Adopted without vote	International Development Strategy for the Third United Nations Development Decade
37/7	20 December 1982	111 (85%9–1–18	World Charter of Nature
37/217	3 December 1986	Adopted without vote	International Co-operation in the Field of Environment
41/65	3 December 1986	Adopted without vote	Principles Relating to Remote Sensing of the Earth from Space
41/128	4 December 1986	146 (94%)–1 – 8	Declaration on the Right to Development
S-18/311 December 1979	1 May 1990	Adopted without vote	Declaration on International Economic Co-operation, in particular the Revitalization of Economic Growth and Development of the Developing Countries
45/19918 December 1979	21 December 1990	Adopted without vote	International Development Strategy for the Fourth United Nations Development Decade

Relevant resolutions of United Nations organs

Resolution	Date of adoption	Voting record	Title
Security Council			
S/Res/330 (1973)	21 March 1973	12 (80%) – 0 – 3	Strengthening of International Peace and Security in Latin America
ECOSOC			
ECOSOC Res. 1737 (LIV)	4 May 1973	20 (77%) – 2 – 4	Permanent Sovereignty over Natural Resources of Developing Countries
ECOSOC Res. 1762 (LIV)	18 May 1973	17 (65) – 0 – 9	Question of the Establishment of a United Nations Revolving Fund for Natural Resources Exploration
ECOSOC Res. 1956 (LIX)	25 July 1975	26 (72%) – 5 – 5	Permanent Sovereignty over Natural Resources
ECOSOC Res. 2120 (LXIII)	4 August 1977	38 (76%)–1–11	Permanent Sovereignty over Natural Resources
ECOSOC Res. 1985/52	25 July 1985	123 (94%) – 0 – 8	Permanent Sovereignty over Natural Resources
ECOSOC Res. 1987/12	26 May 1987	Adopted without vote	Permanent Sovereignty over Natural Resources
ECOSOC Res. 1989/10	22 May 1989	Adopted without vote	Permanent Sovereignty over Natural Resources
ECOSOC Res. 1991/88	26 July 1991	Adopted without vote	Permanent Sovereignty over Natural Resources
UNCTAD			
UNCTAD I	16 June 1964	94 (81%) – 4–18	General Principle Three of the Final Document UNCTAD I
UNCTAD III Res. 46 (III)	18 May 1972	72 (70%)–15–18	Principles Governing International Trade (Principles II and XI)
TDB Res. 88 (XII)	19 October 1972	39 (61%) – 2 – 23	Permanent Sovereignty over Natural Resources
UNCTAD IV Res. 93 (IV)	30 May 1976	Adopted without vote	Integrated Programme for Commodities

In the meanwhile, the Law of the Sea (LOS) joined the evolution of natural resources rights. In 1972 Iceland submitted a draft resolution on behalf of twenty-five developing countries, arguing to 'reaffirm' that permanent sovereignty extends over the resources 'found in the sea-bed and the subsoil thereof within their national jurisdiction and in the superjacent waters'. It was quite obvious that this country sought legitimisation from the UN General Assembly for the position it took in its fisheries disputes with the UK and Germany, on which they had instituted proceedings before the ICJ. The draft met with considerable opposition from both Western countries and non-Western 'land and shelf-locked' countries. They considered it inappropriate for the Assembly to prejudge the outcome of the Third UN Conference on the Law of the Sea (UNCLOS III), which was about to be convened.

The failure of the 1958 and 1960 Conferences to reach agreement on the maximum extent of the territorial sea and the establishment of exclusive fishery zones forced States to take matters into their own hands and resulted in a proliferation of divergent State practices. While most Western nations maintained the three-mile limit, Latin American and newly independent African States began to claim much wider territorial seas. Coastal developing countries felt this was necessary to put a halt to the large-scale exploitation by foreign fishing fleets of what they perceived as their fishery resources. The trend towards proclaiming wider territorial seas culminated in the 200-mile territorial sea first claimed by Chile, El Salvador and Panama. It should be noted that not all of these 'territorial sea' claims followed the definition set out in the 1958 Convention and they were often a mixture of 'territorial sea' and 'functional jurisdiction' claims. After the emergence and recognition of concepts such as the continental shelf during the 1940s and 1950s and the exclusive economic zone during the 1970s, it finally became possible to reach agreement on the extent of territorial seas in the 1982 Convention.[140]

The evolution of oil industry: offshore and national oil companies

Offshore development took off during the 1930s and evolved sufficiently in the following decade to be implemented in the Persian Gulf. Surrounded by the world's largest oilfields and possessing a continental shelf shallow enough for drilling, the Persian Gulf attracted keen interest from the world's oil companies as opportunities abounded for the offshore industry. Shell Oil, which was already producing oil from a 1952 offshore

---

140 *See* NICO SCHRIJVER, SOVEREIGNTY OVER NATURAL RESOURCES (1997).

concession granted by the ruler of Qatar, obtained access to Kuwait's off-shore territory. Iran granted a large offshore concession to Indiana Standard's Iranian subsidiary, the Iran Pan American Oil Company (IPAC). The sheikdom of Abu Dhabi awarded a concession for a joint venture between British Petroleum (BP) and France's Compaignie Françoise des Pètroles (CFP). A mixture of competition and necessary joint action emerged in the Persian Gulf offshore, where high capital costs and exceptional technology were required. Eventually, it also positively affected the relations between private and public companies.[141]

Agreements were signed by Iran with non-majors in the late 1950s. The first was an agreement signed in 1957 with AGIP, a subsidiary of Italy's ENI, to set up a 50/50 joint venture company called the Socitti Irano–Italiènne des Pètroles (SIRIP) between AGIP and the National Iranian Oil Company (NIOC) to explore for and produce oil in a sector of Iran's Persian Gulf offshore. It was the first such joint venture signed in an OPEC country, and was followed the next year in 1958 by a side agreement with Amoco to set up the Iran–Pan American Oil Company (IPAC). This time, the competitive danger was not so much the volume of oil that might be produced from these concessions, nor the privileged tax position of the concessionaires (as previously in Libya), but quite the reverse: the concessions appeared to be more favourable to the government in three important respects: first, they provided for a carried interest in favour of NIOC during the exploration period (i.e., NIOC's partner would bear all the costs of exploration until the discovery of oil in commercial quantities, and would only then be reimbursed for NIOC's share); second, the duration of the agreements was much shorter (a maximum of forty years from the start of commercial production) than for the majors' concessions; and third, they incorporated stringent relinquishment provisions. In 1957 Saudi Arabia signed an agreement with the Japan Petroleum Trading Company (JPTC) which covered the Saudi's fifty percent share in the Neutral Zone offshore area, providing the government with a fifty-six percent share in the company's total net earnings, and an option for the government to acquire a ten percent equity interest in the Japan Petroleum Development Corporation (JPDC) at par share value after the discovery of oil in commercial quantities. The following year, Kuwait signed an agreement with the Ara-

---

141 *See* JOSEPH A. PRATT, TYLER PRIEST and CHRISTOPHER J. CASTANEDA, OFFSHORE PIONEERS: BROWN & ROOT AND THE HISTORY OF OFFSHORE OIL AND GAS, 108 (1997.).

bian Oil Company (AOC) covering its fifty percent share of the offshore Neutral Zone (the onshore Neutral Zone was covered by earlier agreements between Saudi Arabia and Getty in 1949 for the Saudi share, and between Kuwait and the American Independent Oil Company (Aminoil) in 1948 for the Kuwaiti share; but these two earlier agreements did not incorporate terms significantly more favourable to the governments than the majors' concessions). The obvious and natural consequence of these agreements was to increase the host government pressure on the major oil companies for the revision of the existing concessions. This was a competitive threat of a different kind, because while it did involve additional volumes of oil in the hands of the non majors, the majors were more apprehensive about the discontent they engendered in the governments over existing terms – any upward revision of the fiscal terms in favour of the government would erode the majors' competitive advantage. Later on, in the late 1960s and early 1970s, there were to be a number of other offshore agreements signed with non-majors, but by that time, fundamental changes in the majors' concessions had either been concluded (with the Tehran Agreement of 1971) or were in train. In the event, the amounts of oil produced under the terms of the Iranian offshore and the offshore and onshore Kuwait/Saudi Arabia Neutral Zone agreements with the non-majors grew to a respectable volume over the years, increasing from 229 thousand barrels per day (kbd) in 1961 to 609 kbd in 1966 and 1,025 kbd in 1970.[142]

National oil companies in the developing world emerged either from nationalisation – taking over the expropriated assets of the foreign oil companies – or from 'participation' agreements in which a national oil company gradually displaced a foreign oil company as the state purchased the company's assets. Within these two types of NOC formation, there were variations or subtypes that distinguish each case. The key factors in NOC formation were whether the NOC had the time and mechanisms needed to learn or retain skills from the foreign oil companies and whether or not the relations with the concessionaires and their home governments were conflictive. The first wave of national oil companies emerged out of nationalisation. Mexico and Iran were ahead of a number of other producers in this process, though foreign oil companies continued to operate in Iran, albeit as contractors to NIOC. Relations between the Shah's govern-

---

142 *See* Parra, *supra* note 19, at 86.

ment and the foreign oil companies were difficult. The Shah constantly pressed the BP-led consortium to increase production, pay out a larger share per barrel and invest in increasing capacity. Indeed, Iran increased its oil supply to the foreign consortium during both the Suez crisis and the Six-Day War. In 1966 the Shah threatened to turn to the Soviet Union for cheaper arms purchases, arguing that Iran's oil revenues were insufficient. The British government encouraged the consortium to settle, and terms were agreed. It increased production, supplied NIOC with discounted oil for bartering outside its members' own markets and reduced the area of the concession. In 1968, after Britain had announced that it would withdraw all its forces from the Gulf, the Shah again demanded more revenue from the consortium, this time with the support of the US and the UK governments. After a further confrontation, the consortium agreed. Foreign oil company activity and the political interference in it continued until the revolution in 1979.

In Kuwait relations between the government and the oil companies became a topic of vigorous public debate during the 1960s. Though the terms of the concession had been renegotiated in 1951, the rising importance of Kuwait in the world oil market by the 1960s had caused some to question the existing arrangements. Oil became the dominant theme of debate during the first session of the Kuwaiti parliament, established in 1963, when a group of opposition MPs criticised the practices of the Kuwait Oil Company (KOC, owned by BP and Gulf Oil) and the failure of the pro-Western government to combat them. In particular, it called attention to the overproduction of reserves, which had damaged the long-term recoverability of the Burgan oilfield; to the practice of flaring the gas by-product associated with the production of oil, rather than putting it to commercial use; to the limited progress made in training and employing Kuwaitis in skilled positions; and to KOC's failure to capture the lucrative market supplying bunker fuel to Persian Gulf oil tanker traffic, which could have been carried out by the state-owned downstream company Kuwait National Petroleum Company (KNPC). A closely followed press campaign brought these issues to the attention of broader Kuwaiti society: and in parliament, the dispute came to a head over the Expensing of Royalties Agreement (established by OPEC in 1965), which renegotiated the fiscal terms for the concessionaires on a basis opposition MPs saw as excessively lenient. The government stood firm on its close relations with the concessionaires, eventually passing the original bill in 1967 after having called fresh elections in which the opposition group had been weakened.

In Iraq the monarchy founded by the British in 1925 was overthrown in 1958 by a *coup* supported by Nasserites and communists. The new government of General Qassim made a series of demands on the IPC, and in December 1961 it passed Law 80, expropriating without compensation all of the IPC's concessions except those fields which were actually producing. After General Qassim was overthrown in 1963, the nationalist Iraqi government granted the Iraqi National Oil Company (INOC) rights to all onshore oil deposits in Iraq except the small area of existing production allowed to the IPC under Law 80. Later the same year, it reorganised INOC and brought it under the direct control of the government. INOC adopted a policy of diversifying markets and development contracts, especially favouring the USSR and France, which supplied arms, aid and technology to the regime. In 1968 the period of political turbulence was ended by a Ba'athist *coup*, after which Iraq was governed by the Revolutionary Command Council. The remaining IPC oilfields were nationalised in 1971 and 1974, with compensation paid as in other OPEC countries, on the basis of the book value of the installations.

The situation in Abu Dhabi was quite different. Owing to difficult geological conditions and a lack of natural harbours, prospecting began there later than elsewhere: the first major discoveries were made onshore in 1960 and offshore in 1962. By this time, a major influx of oil revenues into neighbouring Qatar and Saudi Arabia had already caused a number of locals to migrate in search of jobs and a higher standard of living. Pressure mounted for a renegotiation of the concessions, which looked increasingly outdated, in order to enlarge revenues and fund investments in healthcare, education and infrastructure. In 1965 a new agreement brought royalties and taxation into line with the OPEC norm and reduced the size of the concession area to the producing fields. This arrangement freed the government to explore in the remaining areas or to invite bids from other companies to do so. When in 1966 Sheikh Zayed replaced his brother as ruler of Abu Dhabi, he put pressure on the foreign oil companies to increase exploration and production so that his ambitious modernisation programme could be paid for, but he maintained a cooperative attitude towards them which contrasted with the nationalist confrontations elsewhere in the Middle East.

In 1972 the producers which had not yet nationalised their oil industry were heatedly discussing the relative advantages of participation versus nationalisation. Governments hoped to gain control of the industry in order to resist foreign dominance – this was deeply resented by the population – but they feared the Seven Sisters. They were concerned that a country which nationalised the interests of such powerful international companies would then be excluded from the world oil market in an industry dominated by those few companies. With participation, however, a government took a share of a company's concessions, rights, obligations and operations in its country, which was regarded as an acceptable compromise that avoided the uncertainties resulting from nationalisation. The Participation Agreement, which came into effect on 1 January 1973, was signed by Saudi Arabia and Abu Dhabi and the major oil companies operating in those states. It allowed the producing countries to take a twenty-five percent share in operations, which would rise by five percent each year until it reached fifty-one percent. Compensation would be paid to the companies, but only for the 'adjusted book value' of the facilities, not for the oil resources, and the companies would assist in marketing the oil. Qatar, and later Nigeria, concluded similar agreements with their concessionaires.

However, the end result was in fact a mixture of both participation and nationalisation, because soon after the 1973 agreement, Abu Dhabi negotiated a further agreement with the companies. This took effect in 1974, and in it the government acquired a sixty percent share in operations and the companies retained forty percent. Saudi Arabia and Kuwait, for their part, would nationalise the foreign oil company interests completely: the former would do so over a period of years and the latter rapidly.

When the Kuwaiti government submitted its proposal for an initial twenty-five percent participation, its parliament (supported by public opinion and the press) was hawkish. It pressured the government to demand a greater initial share from the companies and commitments to reduce gas flaring. In 1971 Kuwaiti officials explained to Gulf Oil and British Petroleum that the government wished to avoid a collision with its parliament and that the issue of gas utilisation could therefore not be ignored. The companies' response did not satisfy parliament. They apparently underestimated its influence and that of public opinion on the government. Parliament rejected the Participation Agreement and from 1974, the Kuwaiti government began to nationalise the upstream industry, acquiring a sixty percent participation in the Kuwait Oil Company. This hybrid participation–nationalisation deal came to an abrupt end in March 1975, however, when Kuwait announced that it would take over the remaining forty percent. For its part, the Saudi government gradually bought up the Arabian–American Oil Company (Aramco). It acquired a one hundred percent participation interest by 1980 and in 1988, it established the wholly state-owned Saudi Arabian Oil Company (Saudi Aramco), thereby taking over Aramco's activities.

The creation of national oil companies helped the producing states to nationalise oil by giving them the technical and organisational means gradually to take over operations from the private companies. In most cases, there was a period during which the national oil company and the foreign companies overlapped. As we shall see, the quality and duration of relations between the foreign companies and national oil companies had a significant impact on the national industry later on. This was a period during which the national oil companies could learn the business. In particular, the slow transition of participation agreements allowed states to acquire the title to their resources and to develop the means to become operators of their own oilfields in partnership with foreign oil companies. This process was especially fluid in Saudi Arabia, where Aramco was progressively nationalised. It was a smooth transition whereby the original company's expertise and organisation were largely retained thanks to technical and marketing agreements with the foreign consortium members which lasted until the mid-1980s. Interestingly, there was another national oil company in Saudi Arabia, initially created to develop the Kingdom's mineral resources: this was Petromin, established in 1962 as a public corporation active in the downstream business. However, it failed to develop the skills and expertise necessary to develop into the main NOC, and the Saudi government chose instead to keep Aramco as the upstream operator and to 'buy it back' from the foreign oil companies.

In contrast, the National Iranian Oil Company emerged in the turbulent politics of mid-twentieth-century Iran. NIOC was established in 1948 as a first step in the nationalisation process at a time when Anglo-Iranian still held concessions for most of Iran's known oil. Between nationalisation in 1951 and the departure of the last element of the international consortium in 1979, the role of the foreign companies in Iran was repeatedly challenged and curtailed.

In 1973 the Shah negotiated terms for a new twenty-year agreement: this was predicated on the desire to obtain the financial equivalent of the twenty-five percent participation originally proposed by Saudi Arabia. When the Saudis gained more favourable terms, the Shah increased his demands.

The consortium operated the oilfields, and the Shah wanted more control over the industry. He got what he wanted: NIOC took over the consortium's assets and the consortium, as the Oil Service Company, continued to manage operations and to pump oil under a five-year contract. In this context relations between the national oil company and the foreign operators were not conducive to a transfer of skills to the NOC. Finally, after the revolution of 1979, the new Ministry of Petroleum cancelled all existing oil agreements and took control of oil and gas operations through the NIOC, the National Iranian Gas Company (established in 1965) and the National Petrochemicals Company (established in 1964). By the end of 1979, NIOC had set up the National Iranian Drilling Company, which began drilling and also maintaining twenty-seven abandoned rigs.

Like Saudi Arabia, Kuwait initially created national companies which only participated in the non-core areas of the oil sector alongside the foreign companies which were active in the core upstream activities. Between 1957 and 1963, Kuwait founded the Kuwait Oil Tankers Company, the Kuwait National Petroleum Company and the Petroleum Industries Company as joint stock companies with private shareholders. The core exploration and production company, the Kuwait Oil Company (KOC), established in 1934 by foreign oil companies, was acquired by the Kuwaiti government in two parts, in 1974 and 1975. Because of the political pressure exerted by the Kuwaiti parliament, the government indicated that it did not want to maintain special links with BP and Gulf Oil. In response BP and Gulf Oil representatives went to Kuwait City for talks and argued for consideration to be shown for the old relationship. However, the Kuwaitis told them emphatically that this would not happen and that Kuwait intended to take over one hundred percent, that it was a matter of sovereignty, and that the question was not open to debate. The concessionaires claimed USD 2 billion in compensation but were only given USD 50 million.

The Abu Dhabi National Oil Company (ADNOC) was established in 1971 to engage in all activities in the hydrocarbon sector It would have a number of affiliates and subsidiaries. Following the Participation Agreement, it took on the state's interest in the concessionaires' rights, obligations and operations. Unlike most Gulf countries, Abu Dhabi never claimed one hundred percent ownership of the industry. In 1974 it took a sixty percent participation in onshore and offshore concessions. Today, most production is still carried out by the pre-1974 consortia in which ADNOC is the majority partner, though it does explore for new concessions and may undertake developments on its own initiative. The original tax structure was retained, with the foreign companies paying royalties and taxes (under terms less favourable than before 1974: effectively, they receive a fixed margin of approximately USD 1 per barrel produced rather than a share of the profits).

Countries which nationalised or negotiated the purchase of foreign oil company assets created different conditions for the creation of their national oil company. Of those which nationalised their industry, some created a national oil company well ahead of the expropriation of upstream assets. This was the case in Algeria, which as indicated, created Sonatrach initially to control the pipelines. Other assets were nationalised over a period of years, which gave the country time to build up its managerial and technical capacity. For instance in Iran, state ownership of resources was entrusted to NIOC, which oversaw the work of foreign oil companies until complete nationalisation. Kuwaiti companies operated in Kuwait's non-core activities from 1957, but the foreign oil companies operating upstream alongside KOC coexisted for only a short time with the national staff before full nationalisation in 1974–1975. Furthermore, after 1975 KOC did not retain foreign oil company personnel on its board or in its operations.

In these cases of expropriation, the national oil companies were created in a context of conflict with the concessionaires or the concessionaires' home countries. Although Sonatrach and NIOC coexisted with the foreign oil companies for some years before their departure, the difficult history of political emancipation from foreign interference (whether plotting and *coups* in Iran or war in Algeria) did not create the conditions for skills transfer and the joint development of resources. After full nationalisation in Algeria in 1971 and the revolution in Iran, the national oil companies operated largely on their own. Iran and Algeria were deeply affected by their difficult relations with Britain and France respectively. Their NOCs have tended to be more politicised and their societies have also been more sensitive to foreign investment. In Kuwait the population was also mobilised against foreign investment, but events played out differently because their rulers, who were seen as pro-Western and too soft on the concessionaires, remained in place (unlike Iran). As a result, the tensions were greater between the government and parliament than between the government and the concessionaires. Kuwait also differed from Algeria and Iran in that its relations with Britain were relatively peaceful, even though those with the concessionaires ended on a sour note. In Abu Dhabi and Saudi Arabia, relations with Britain and the United States were also positive, and this could explain why those national industries are less politicised than those in Iran and Algeria. The historical conflict between the Kuwaiti government and its parliament certainly explains why members of parliament remain watchful of the government's management of the oil sector.

For countries such as Saudi Arabia and Abu Dhabi, which decided on participation agreements and a gradual nationalisation of the oil industry, there were greater opportunities to learn skills. ADNOC in particular still operates with the original concessionaires through partnerships with them in most of its operating companies. Saudi Aramco also maintains positive relations with the concessionaires. For instance, its board still includes former CEOs of foreign oil companies. As a result, these national oil companies developed strong managerial and operational processes at an early stage. Even where there were participation agreements, however, governments throughout the Middle East were ill-informed up to the 1970s about even the most basic facts required to enable them to formulate their oil and financial policies. Although oil companies were compelled to provide the necessary information to governments about their operations, they did so at a trickle during the early years of participation. Governments and national oil companies had no direct access to the data on their reserves from the companies, and relied on secondary sources or the industry press. For all the vast amounts of raw data that were transmitted from the companies to the governments during the early years of participation, there was a total embargo on 'interpretative' data. Producers came to mistrust the foreign oil companies, seeing them as secretive and arrogant.

It goes without saying that nationalisation was essential to support the Middle Eastern elites' regimes and legitimacy. Governance over the oil and gas sectors was a high priority for governments and national oil companies in the Middle East. Accordingly, companies were directed from the outset by a Supreme Petroleum Council (or equivalent) formed of government representatives – the head of state, financial authorities and ministers (usually presided over by the Oil Minister) – and company representatives.[143]

---

143 *See* VALÉRIE MARCEL, *supra* note 13, at 76.

Conclusions

This analysis suggests various reasonable sources for the institutional design of the seven treaties under consideration, particularly concerning their brevity, narrow subject matter and specific though programmatic terms.

To begin with, we must consider their interaction with the worldwide political economy – an enlightened treaty-craft prevailed in the Middle East between 1958 and 1974. The States which drafted and became parties to them had to overcome Cold War tendencies as well as steer clear of regional conflicts in order to achieve and sustain their distinctive economic positions globally. Beyond the form that their discourse might take, it was important for these States to *proceed* in a liberal manner: overall, the core aim was the securing of oil provision to Western countries and obtaining better revenues. By the time the treaties were negotiated and signed (1958–1974), the states parties' movement towards political economic liberalism had already begun. The only thing left to achieve for a fully stable governmental position was to complete the process of taking control over the natural resources.

Strange as it may sound, I view the elites' progressive national consolidation as a 'symbiotic' form of chronological interaction between the factors involved. First, there were semi-authoritarian regimes supported by nationalist discourses and 'clientelism' based on high oil revenues. Then, public international law evolved towards absolute national sovereignty over natural resources, and in the meanwhile the international LOS took clearer shape, inducing the expansion of state sovereign rights. Finally, there was a substantial increase in the states' revenues from the nationalisation of the oil industry and offshore oil development, enabled by technological improvements. In my opinion, this 'symbiosis' characterised the treaties' design. There was a possibility to expand material gains and to put an end to long-lasting territorial conflicts: with that, legitimacy could be further assured.

Institutional cooperation: warfare and economy (1973–2010)

There were several reasons for the failure to sustain permanently high oil prices after 1973. Increased conservation in the importing states was one factor. Another was that OPEC states encountered policy obstacles, the limits of political economy, both internationally and internally. They found it difficult to convert their market control of the early 1970s into a permanent international lever; they found it even harder to use their natural wealth to further broader foreign policy goals. Those states, notably Saudi Arabia, which sought to use their oil wealth to gain influence in the Arab world, found it to be an unreliable 'weapon'.[144]

In the 1980s divisions developed within OPEC between population-rich countries (e.g., Iran and Iraq), which wanted to maximise output, and low population states (e.g., Kuwait and Abu Dhabi), which wanted to conserve oil resources. Oil pervaded the economies of Middle Eastern producer and non producer countries alike, but in doing so it exacerbated existing weaknesses and hierarchies as much as it reduced them. With regard to the demand for sustained regional growth, oil did not integrate the economies of the Middle East except in terms of financial flows. Rather, it reinforced the fragmentation and suspicious bitterness of producer states. Ultimately, the common possession of oil did not produce a common foreign policy: Iran quarrelled with the Arab world, and Iraq with Kuwait.[145]

On the other hand, if the 'Oil Weapon' had limitations when applied to the Middle East, this was even more evident in relations with the outside world.[146] During the crisis of 1973 following the Arab–Israeli war, the Arab producers reduced output by twenty-five percent and vowed not to supply the US or the Netherlands, the two states considered most sympathetic to Israel, until Palestinian demands had been met. In the end, and despite much alarm in the West at Arab 'blackmail' (as if all market relations were not based on conflicting pressures), this concerted Arab action

---

144 Egypt, whose break with the Soviet Union was in part financed by the Saudis in the early 1970s, defied Saudi advice in making peace with Israel in 1977–1979. Yemen, itself without significant resources, was more antagonised than reconciled by Saudi supplies of funds to state and tribal leaders alike. In 1990–1991 many of the Islamic militants who had until then been funded by Saudi Arabia defied Riyadh and supported Iraq in the confrontation over Kuwait. All the recipients of Kuwaiti international political economy funding before 1990 – Sudan, Yemen and the Palestinians – also sided with Iraq in that conflict.

145 For an evolution of Saudi Arabia's unilateral economic relations, *see* GERD NONNEMAN, *Saudi–European Relations 1902–2001: A Pragmatic Quest for Relative Autonomy*, 3 International Affairs (2001) 631; for a detailed evolution of Iraqi–Kuwaiti foreign policy relations, *see* MICHAEL S. CASEY, THE HISTORY OF KUWAIT (2007).

146 *See* Parra, *supra* note 19, at 139.

achieved nothing. In early 1974 the boycott ended: not a single positive political result for the 'Arab cause' resulted then or during the rest of the century. Many outside the Middle East seemed to think that Western policy faced a conflict between its inclination to support Israel on the one hand, and its relations with the Arab oil producers on the other – this was a concern often voiced by those sympathetic to Israel itself. However, this dilemma was largely illusory, in that on closer examination, Western relations with the oil producers were not at all affected by the Israeli connection.[147]

It may be useful to insert a brief parenthesis here to explain that the consequences in international relations of the presence of oil 'as such' were not far-reaching. Despite much rhetoric to the contrary, 'oil' did not promote conflict any more than does water or wheat, or frontiers or religion for that matter. The impact of oil in international relations, as in internal relations, depends on the policies of states and of those who challenge them. It was nationalism and social conflict, driven by the power calculations of state elites and their rivals, which turned oil into controversy. Resentment at external or oligarchic control of oil and oil rents was a recurrent feature of Middle Eastern politics. So too was conflict in which a threat to oilfields and wealth was perceived.[148]

Now, as discussed above, Middle Eastern states in many ways became more not less vulnerable after 1975: later events would demonstrate how far this vulnerability would go. After 1973 it was not in the exacerbation of inter-state relations that oil was so important, but in the sharpening of state–society relations within states. The revolution in Iran of 1978–1979, the rise of Islamic fundamentalism in Algeria in the late 1980s, and the 1991 uprising against Saddam Hussein in Iraq, upheavals in three large 'rentier' states, showed how the misallocation of oil revenues could fuel social tensions which ultimately challenged apparently strong states. However, the Iraqi invasion of Kuwait in 1990, following which Kuwait and

---

147 Moreover, the 'Oil Weapon' itself was never used or seriously contemplated again after 1973. Indeed, close on two decades were to pass before, for different reasons, an Israeli–Palestinian compromise was to be worked out at Oslo in 1993. After that, Arab economics and their financial support for the PLO, eroded by Palestinian backing for Iraq in 1990–1991, had no impact on the peace negotiations or on the conflict.

148 *See* TAHIR HUSAIN, KUWAITI OIL FIRES: REGIONAL ENVIRONMENTAL PERSPECTIVES (1995).

Saudi Arabia were forced to call for international help, showed how little oil rent could be converted into military security. Despite the shift in ownership and apparent market influence which began in the early 1970s, the oil states were not therefore more powerful in foreign policy or strategic terms. Politics and economics were interwoven, but strength in one was not necessarily convertible or, to use the economics term, 'fungible' into strength in the other. This was the point of departure of the close relationship between the United States and Saudi Arabia. Aware of its security weakness, the latter accelerated its ties with the United States to exchange this scarce, essential, good for oil.[149]

After decades of speculation about oil, oil revenues had two most tangible results. One was the increased import of arms into the region and the consequent reduction in security, domestic and international, that the importing states felt.[150] The other was the concern showed by the industrialised states, and in particular the US, to ensure reliable and uninterrupted access to the oil of the region. Neither did much for the long-term social and economic development of the region. At the same time, while OPEC held 73.5 percent of proven world reserves in 1998, the rise of non-OPEC producers such as Mexico, Norway, Russia and Colombia, cause OPEC's percentage of the world market to fall from its height in the early 1970s to forty-two percent in 1998. In the 1990s technological change also significantly reduced the role of energy in economic growth. Politics and state

---

149 *See* WAYNE H. BOWEN, THE HISTORY OF SAUDI ARABIA (2008); also RACHEL BRONSON, THICKER THAN OIL: AMERICA'S UNEASY PARTNERSHIP WITH SAUDI ARABIA (2006). In this line, there is an interesting study of the Saudi's custom of reserving a 'second best option'. Before the Second World War, the options were the Axis or the Allies, with prevalence of the later. With the arrival of the Cold War, the options became the United States or Europe, *see* Gerd Nonneman, *Saudi–European Relations 1902–2001: A Pragmatic Quest for Relative Autonomy*, 77 INTERNATIONAL AFFAIRS 631–661. In the future, Rachel Bronson states that an eventual partner could be China, if the US steps back.
150 The possession of oil was on occasion itself a source of particular anxiety, evident in the 1960s in Saudi fears of Egyptian advances through Yemen, through to the Iran–Iraq war that broke out in 1980 and the Iraqi invasion of Kuwait in 1990. However, such anxiety, related to the protection of an immovable resource, combined with the fear of pressure from within, led these states to engage in substantial arms purchases and hence to a regional arms race, which grew from the early 1970s, in parallel to the rise in oil revenues.

intervention by OECD members were also evident in the downstream sector.[151]

In regional economic terms, modern politics had, indeed, as much divided as united the region. Regional trade was inhibited by 'modern-state' institutionalisation, just as it was by fractured political and cultural contact between peoples and tribes. In part this might stem also from the belief that trade with more developed and powerful states may have other benefits for the Middle Eastern state concerned: an obvious example is in US–Saudi relations, where closer cooperation is the guiding principle to ensure defence security. This politicisation of trade to exclude regional economies has been taken further by the stress on import-substitution industrialisation. For all the talk of Arab, or sub-regional economic integration, in the early 1990s intra-regional trade accounted for only ten percent of the total.[152] This limited regional trade was, however, compounded in the Middle East by the fact that non-oil exports from these countries were in the main relatively low: that is, they had relatively little to trade with each other. Again, the 'politics' of integration prevailed over economic criteria. In the words of the noted Iraqi economist Mohammad Salman Hassan: 'In the Arab world there are no state-to-state economic agreements, only person-to-person ones'.[153] Individual, sovereign, rulers gave loans to rulers in other states on the basis of individual interest – and sometimes, trust. Arab states, or more accurately Arab rulers, invested and loaned for political purposes, supporting regimes they favoured or wanted to influence. They denied support, or withdrew it, when they did not favour the recipients.

The story of economic integration programmes and of the creation of an Arab common market is itself a limited one, with projects for economic cooperation or integration being prompted and then paralysed or cancelled for political purposes. Thus Gulf Air, established by the smaller Gulf Cooperation Countries states in 1980, contracted after a few years as individual states founded their own carriers – Qatar Airways, Oman Air and Emirates. In the late 1980s three different political and economic groups were apparently in operation within the Arab world – the Gulf Coopera-

---

151 *See* Halliday, *supra* note 9, at 153.
152 *See* BADR EL DIN A. IBRAHIN, ECONOMIC COOPERATION IN THE ARAB GULF: ISSUES IN THE ECONOMIES OF THE ARAB GULF COOPERATION COUNCIL STATES (2007).
153 *See* Halliday, *supra* note 9, at 281, n. 46.

tion Council (GCC), the Arab Maghrib Union and the Arab Cooperation Council. The GCC had been founded in early 1981 for political and security reasons, after the outbreak of war between Iran and Iraq the previous September. The Arab Maghrib Union involved a low-level set of economic arrangements. The Arab Cooperation Council, comprising Egypt, Jordan, Iraq and Yemen, broke apart in 1990 when Iraq sought confrontation with its Arab neighbours and invaded Kuwait. For political and economic reasons, the prevalence of a pan-Arab sentiment in the popular mind, coupled with continuing flows of remittances and funds, was not accompanied by any sustained, let alone institutionalised, integration of the Arab economies. State interests and the allocation on general grounds of security and profitability of resources in a global financial and investment climate, prevailed. Here, much was made of 'Western' inhibitions of Arab economic development but the choices made were those of the regional rulers and investors.[154]

The appearance of growth was in several respects deceptive. First, while overall growth in the Middle East and North Africa in 1962–1975 was 4.91 percent, this fell to an annual average of 0.75 percent from 1975 to 1990 and reached −0.3 percent in the period 1990–1995, in both periods the lowest in the world. The large rises in oil revenues, often in countries with small populations, did not translate into large per capita incomes for the region as a whole. If these absolute figures are corrected for population growth to yield figures for GNP per capita growth, then we get a figure of −0.47 percent for the period 1962–1990. This comparatively poor record for the region as a whole underlines many of the difficulties which economic development faced, not only in terms of the distribution of natural endowments, but also in terms of the impact on economics of the range of political factors characterising the region – wars and arms races, state interventions in and distortions of the economy, the persistent discouragement of entrepreneurial activity, generally poor levels of administrative and educational competence, and the lack of a trained, internationally competitive labour force. Nevertheless, the availability of large sums of money through oil revenues and foreign investment, however unevenly distributed, and the provision of other forms of rent for security reasons,

---

154 *See* Badr El Din A. Ibrahin, *supra* note 36, at 135.

did contribute to an element of political and regime stability in many countries.[155]

Estimates for the period 2000–2025 suggest, indeed, an even greater world reliance on the oil produced by the core Gulf countries – Iraq, Kuwait, Saudi Arabia and Abu Dhabi – than has hitherto been the case. However, other developments, in train since the 1970s and later subsumed into the catchall term 'globalisation', suggest that the Middle East could, and on current performance will, become even more marginal to the world economy than has been previously. On the basis of intergovernmental and international financial institutional criteria and despite an unquenchable flow of grovelling and evasive reports on one kind of 'transition' or another by extra-regional consultants, the Middle East has failed on many counts to meet the criteria of privatisation, liberalisation of domestic markets from subsidies, good governance and transparency which are now all held to be conditions for sustained support and aid, as well as being preconditions for growth. While Middle Eastern states did often take some formal steps to meet such criteria, they were caught in the horns of a set of political dilemmas to which they have only sought short-term solutions: insofar as they implemented the conditions of the World Bank and the IMF or other comparable institutions which followed their criteria, and, for example, cut subsidies, shed workers or indeed contemplated introducing taxes, they provoked political discontent at home. These conditions required that state support for employment and subsidies for prices or social services be reduced. Insofar as the states resisted, or merely appeared to comply, the support they received was reduced. In effect, the states stalled as best they could and used the IMF and other external financial support as a form of rent. However, confronted by external pressure to change on one side, and rising a domestic preoccupation with employment and long term trends on the other, the regimes' room for manoeuvre narrowed, hostage above all to the vagaries of oil prices. It was indeed evident that in much of the Arab world this cooperation, with the increasingly exigent criteria of global financial credibility, was often half-hearted and inadequate: subsidies were cut and agreements on 'restructuring' were signed, but neither a reduction of the state's role in the economy nor increased accountability to any significant degree was to mark Middle

---

155 ALI R. ABOOTALEBI, *Middle East Economies: a Survey of Current Problems and Issues,*
3 Middle East Review of International Affairs (1999).

Eastern economies in the 1990s.[156] As for credible public figures on state finance, there was little hope.[157]

In terms of private economic relations, the picture was even less favourable. This was most obviously the case with regard to flows of private capital through private bank credit, foreign direct investment or portfolio investment: as already noted, while the total, global figure for net private flows to developing or emerging countries rose in the mid-1990s to USD 200 billion, the Middle East, Israel excepted, attracted only a few billion. This limited FDI reflected the broader marginalisation of the Middle East within the new international economic and financial system: the region did not produce significant amounts of goods for export, nor did it attract investment to produce goods there as a result of a range of negative factors, from political instability and lack of good governance to the low level of education of the labour force. The response of many in the Middle East was, moreover, to compound this by their own actions, in particular by capital flight. It is estimated that by the end of the 1990s, over USD 1,500 billion – perhaps over USD 4,000 billion – in regional private capital was invested outside the region (US GDP was, in comparison, USD 10,000 billion). This, a reverse flow of capital, was the main form of regional participation in global financial markets and longer-term business.[158]

As bad as the picture might be at the state level, the reality was different at oil industry level. Reluctant as Saudi Arabia and Kuwait were to give equity access to their reserves, in the end their NOCs would open up their projects to international oil companies (IOCs) and other NOCs, particularly those which concerned offshore engineering.[159] Middle Eastern

---

156 *Id.*
157 *See* DODGE, TOBY AND HIGGOT, RICHARD, GLOBALIZATION AND THE MIDDLE EAST: ISLAM, ECONOMY, SOCIETY AND POLITICS (2002).
158 *See* Middle East Economic Digest (January 2002).
159 Overall, they would seek to maintain control over the management of their resources and to maximise the 'government take'. Agreement could more easily be reached on returns, but investment models were not sufficiently flexible to address other producer concerns. A crucial problem with partnership models was that the state would control the management of reservoirs without sufficient incentives for optimising the development of the host country's resources. Moreover, conventional partnership models did not make full use of the emerging trends transforming the industry to develop new strategic relationships. The terms

NOCs generally distrusted other NOCs in the region almost as much as they did IOCs – though for different reasons. In the case of high-capacity NOCs such as Saudi Aramco, distrust was tied to disdain for the capacity of their national counterparts. With medium-capacity NOCs, there was essentially a lack of interest in the other NOCs' assets. Middle Eastern NOCs prefer to deal with the best available investors, and they were the super-majors and their traditional partners.[160] Nonetheless, projects have been planned and executed offshore for fifty years, without regard to the legal decisions of the state governments.[161]

This section should close with a reflection on a present-day reality and an assertion of a trend. It is a fact that four out of top ten exporting countries in the world were from the Middle East in 2008 (Saudi Arabia, Iran, UAE and Kuwait).[162] However, as some authors affirm, '[a]lthough there is plenty of room for OPEC to influence the oil price in the current oil pricing system, this influence is not unconstrained. [...] [T]he recent changes in the international pricing system have diminished OPEC pricing power, especially when compared to the previous administered oil pricing system. [...] OPEC pricing system is not constant and varies according to oil market conditions. Finally [...] the proposition that OPEC in general

---

of investment were not designed effectively to facilitate joint development and application of technology. The buybacks in Iran or the joint study agreements in Kuwait disappointed their partners in this respect. In the buybacks, the IOCs worked with Iranian companies and thus transferred technology; but they were not contractually responsible for teaching skills and technology. *See* VALÉRIE MARCEL, *supra* note 13, at 209. For a summary of the trends in NOC–IOC offshore relations, *see* Alexandre Oliveira, *Innovation urged in NOC, IOC relations with supplier,* OIL AND GAS JOURNAL (Week of August 3, 2009).

160 *Id*, at 218.

161 *See QATAR–Oil & Gas Fields–The Offshore Fields*, APS Review Gas Market Trends (2003), http://www.allbusiness.com/mining/oil-gas-extraction-crude-petro leum-natural/635460-1.html; *also*, *KUWAIT–The Offshore Fields*, APS Review Gas Market Trends (2005), http://www.allbusiness.com/mining/oil-gas-extraction -crude-petroleum-natural/437634-1.html; Edward Burton, *Saudi Arabia aims to bolster offshore oil and gas program,* Offshore (2009), http://www.allbusiness.co m/mining-extraction/oil-gas-exploration-extraction-oil-oil/13167527-1.html; *Qatar's Offshore Oil & Gas Fields*, APS Review Gas Market Trends (2009), http://www.allbusiness.com/mining-extraction/oil-gas-exploration-extraction-oil- oil/12859086-1.html.

162 Key World Energy Statistics *2008*, INTERNATIONAL ENERGY AGENCY (2009) 11. The other six countries are Russia, Nigeria, Norway, Mexico, Canada, and Venezuela.

and the Middle East in particular are bound to have greater influence in the oil market as they develop their reserves and gain greater share of the market' is not true.[163]

Conclusions

Oil has been inherent to much inter-Arab tension from the 1960s onwards, and it promises to continue to be so in the future: this is the antagonistic perception in the Arab world of the difference between oil-rich and non-oil states, a feeling present in public attitudes in oil-rich and non-oil states alike, long marked by nationalist hostility to the West over its exploitation of Middle Eastern oil, a major factor in Arab politics. Those in the states with oil, and particularly those such as the GCC states with small populations, perceive themselves as under pressure, if not under siege, from their poorer and more numerous fellow Arabs. Here again, the rhetoric of Arab solidarity, of viewing other states as 'brotherly', is limited to discourse and not matched in reality.

The legal path has generally been frozen since the moment the legal step was taken. In some cases, agreements were not only kept unexecuted, but have since been contradicted.[164] States have also been extremely cautious during negotiations to avoid legally binding statements or documents.[165] Finally, the lack of a multilateral organisation with executive mandate has been counterproductive to negotiations.[166]

The proper manner to end this chapter is to provide a possible answer to the original question – was cooperation ultimately undermined by regional warfare and international political economy, or were the treaties dead let-

---

163 Bassam Fattouh, *OPEC pricing Power: The Need for a New Perspective*, in THE NEW ENERGY PARADIGM (Dieter Helm ed., 2007) 280.

164 *See* the exchange of notes between Saudi Arabia and the UAE concerning the Joint Minutes on the land and maritime boundaries to the Agreement of 4 December 1965 between the State of Qatar and the Kingdom of Saudi Arabia on the delimitation of the offshore and land boundaries. Available at http://www.un.org/Depts/los/LEGISLATIONANDTREATIES/STATEFILES/SAU.htm.

165 *See* CASE CONCERNING MARITIME IDELIMITATION AND TERRITORIAL QUESTIONS (Qatar v. Bahrain), 1994 ICJ (Jurisdiction and Admissibility) (July 1.).

166 *See* Gwenn Okruhlik and Patrick J. Conge, *The Politics of Border Disputes: On the Arabian Peninsula*, 54 Int'l Jour. 230.

ters from the outset? I incline myself towards the latter option, as the elites who concluded the treaties never needed legal solutions to these factual problems – those problems were always solved in practice by their own companies, located 'one level below'.

5 The Gulf of Mexico: a two-step crafting of cooperation regime

This chapter will consider an exemplifying case of disjointed craftsmanship in cooperation regimes: the Gulf of Mexico and the bilateral relationship between Mexico and the United States.[167] The main argument in the chapter draws on two theoretical approaches – functionalism and constructivism – to uncover the treaty precedents, institutional design and outcomes.

According to the functional theory of international institutions, institutional design varies with the type and seriousness of the cooperation problem being addressed.[168] In contrast, the basic proposition of a constructivist or sociological-institutionalist alternative explanation is that the design of international institutions varies with the collective identities and norms of the international community which establishes them,[169] and with the requirements of community-building and community representation.[170]

167 *See* Treaty to Resolve Pending Boundary Differences and Maintain the Rio Grande and Colorado River as the International Boundary, 23 November 1970 (entry into force: 18 April 1972; registration number: 11873; registration date: 17 July 1972); Maritime Boundaries Agreement Effected by Exchange of Notes between the United States of America and Mexico 24 November 1976; Treaty on maritime boundaries between the United States of America and the United Mexican States (Caribbean Sea and Pacific Ocean), 4 May 1978 (entry into force: 13 November 1997; registration number: 37399; registration date: 12 April 2001); Treaty between the Government of the United States of America and the Government of the United Mexican States on the Delimitation of the Continental Shelf in the Western Gulf of Mexico beyond 200 Nautical Miles, 9 June 2000 (entry into force: 17 January 2001; registration number: 37400; registration date: 12 April 2001).

168 *See* for example George W. Downs, David M. Rocke, and Peter N. Barsoom, Managing the Evolution of Multilateralism, 52 INT'L ORG 397–419 (1998.).

169 Christian Reus-Smit, *The Constitutional Structure of International Society and the Nature of Fundamental Institutions*, 51 INT'L ORG 569 (1997).

170 Kenneth W. Abbott and Duncan Snidal, *Why States Act Through Formal International Organizations*, 42 JOURNAL OF CONFLICT RESOLUTION 24 (1998).

The Gulf of Mexico's Western Gap TMDCS

Historically, negotiations on land boundary issues between the United States and Mexico have been sensitive and strained, overshadowing the negotiations of the maritime boundary areas between these two countries.[171]

It was not until the 1970s, over 120 years after the Treaty of Guadalupe-Hidalgo was concluded, that both the United States and Mexico began legislative efforts to establish maritime jurisdictional boundaries in the Gulf of Mexico. These negotiations eventually led to the Treaty on Marine Boundaries (hereinafter, TMB) in 1976,[172] before the Treaty with Mexico on the Delimitation of the Continental Shelf (hereinafter, TDMCS) was finally signed and ratified in 2000.[173]

Preliminaries

As Feldman and Colson recall, during the Carter administration '[t]he United States claim[ed] a territorial sea of 3 nautical miles in breadth…a fishery conservation zone of 2000 nautical miles in breadth, and sovereign rights for the purpose of exploring and exploiting the resources of the continental shelf'.[174] In the meanwhile, in 1976 Mexico became one of the first countries to establish a 200 nautical mile exclusive economic zone (EEZ) demarcating its territorial jurisdiction, achieved by amending Article 27 of the Mexican Constitution during President Echeverria's administration.[175] The text inserted after paragraph seven provides:

---

171 *See* Jorge A. Vargas, *Mexico's Legal Regime over its Marine Spaces: A proposal for the Delimitation of the Continental Shelf in the Deepest Part of the Gulf of Mexico*, 26 U. MIAMI INTER-AM. L. REV, 189, 238 (1995).

172 Treaty on Marine Boundaries Between the United States of America and the United Mexican States, May 4, 2978, US–Mexico, 17 L.L.M 1073.

173 *See* Treaty Between the Government of the United States of America and the Government of the United Mexican States on the Delimitation of the Continental Shelf in the Western Gulf of Mexico Beyond 200 Nautical Miles, June 9, 2000, US–Mexico, 8 TREATY DOC. No. 106-39 (2000).

174 Mark B. Feldman and David Colson, *The Maritime Boundaries of the United States*, 75, AM. J. INT'L. 729, 730 (1981.).

175 Jorge A. Vargas, *U.S. Marine Scientific Research Activities Offshore Mexico: An Evaluation of Mexico's Recent Regulatory Legal Framework*, 24 DENV. J. INT'L L. & POL'Y I, 36 (1995.).

The Nation will exercise control over an area situated outside the territorial seas and adjacent to them, under the rights of sovereignty and the jurisdiction that the laws of the Congress determine. The exclusive economic zone will extend to two hundred nautical miles from where the territorial seas start. In those cases in which this extension produces conflict with the exclusive economic zones of other countries, the boundaries of these zones will be determined by means of agreements with those countries.

The creation of preliminary maritime zones between Mexico and the Unites States prompted further negotiations of boundaries between the two countries, leading to the Exchange of Notes on November 24, 1976. Pursuant to the Exchange of Notes, the maritime boundaries were set at 200 nautical miles offshore of each respective coastline. Because the boundaries established in the Exchange of Notes were merely 'provisional', further agreement was necessary. As a result, the Unites States and Mexico completed a formal agreement in 1978, the TMB, regulating boundaries in the Gulf of Mexico (see Map 3.1)

GULF OF MEXICO & CARIBBEAN SEA

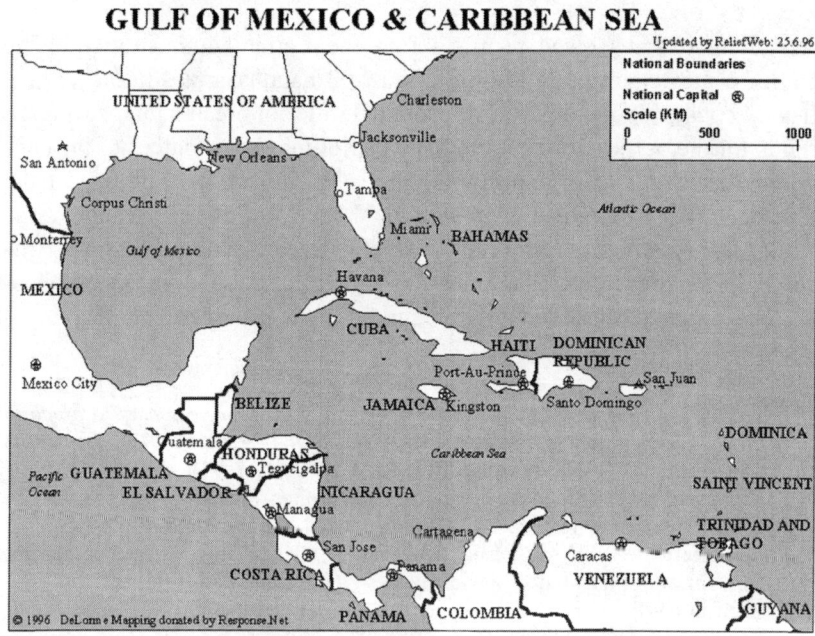

The boundaries and names shown on this map do not imply official endorsement or acceptance by the United Nations or ReliefWeb.
These maps may be freely distributed. If more current information is available, please update the maps and return them to ReliefWeb for posting.

Map 3.1: Gulf of Mexico

Routes to policy convergence (I): US treaty ratification

As late as September 16, 1980, the US Senate postponed consideration of the treaty with Mexico when certain questions arose concerning the presence of oil and gas deposits in the deepest part of the Gulf. As one commentator put it, in an attempt to describe US concerns: '[b]ecause of the serious and permanent consequences that derive from this act of national sovereignty, the establishment of national boundaries is one of the most important decisions a nation can make under international law'.[176] No further action was taken to ratify the TMB until the mid 1990s, when opposing opinions from academics and the US Geological Survey regarding the magnitude of true oil reserves were settled. In addition, industry was ready: technological advances had enabled deepwater exploration in the deepest parts of the Gulf of Mexico, at over 10,000 feet depth, and six domestic energy groups urged prompt ratification.[177] Finally, the United States had reached the point where it needed to exploit the reservoir, as daily crude oil imports had risen to above fifty percent, employment was in decline and there was the prospect of the longer-term improvement in the environmental control of the area.[178] The US Senate finally ratified the TMB on October 23, 1997, and the Treaty entered into force on November 13, 1997.[179]

Once the US Senate had ratified the TMB, the maritime jurisdictional boundary lines were extended to 2000 nautical miles off of each country's respective coastlines. Because the boundary lines did not overlap in some areas, the Western Gap, an area of 4.5 million acres of seabed roughly the size of the state of Iowa, was left undivided by the TMB (*see* Image 3.1). The Western Gap is located between Texas and Mexico's Yucatan Peninsula. The economic significance of this area for petroleum exploration was great, as geologists believed that it could be the world's fourth largest oilfield. The unexplored sources of hydrocarbons and natural gas in the deepest part of the Gulf of Mexico – beyond the maritime boundary regions –

---

176 *See* Vargas, *supra* note 1, at 490.
177 *Id.*, at 459–460.
178 David B. Sheinbein, *Delimitation of Western Gap Land in the Gulf of Mexico: A Need for Diplomatic Resolution*, 6 TUL. J. INT'L & COMP. L. 583, 595.
179 *Id.*, at 463.

contain between 2.224 billion to 21.990 billion barrels of oil and 5.48 trillion to 44.40 trillion cubic feet of gas.[180]

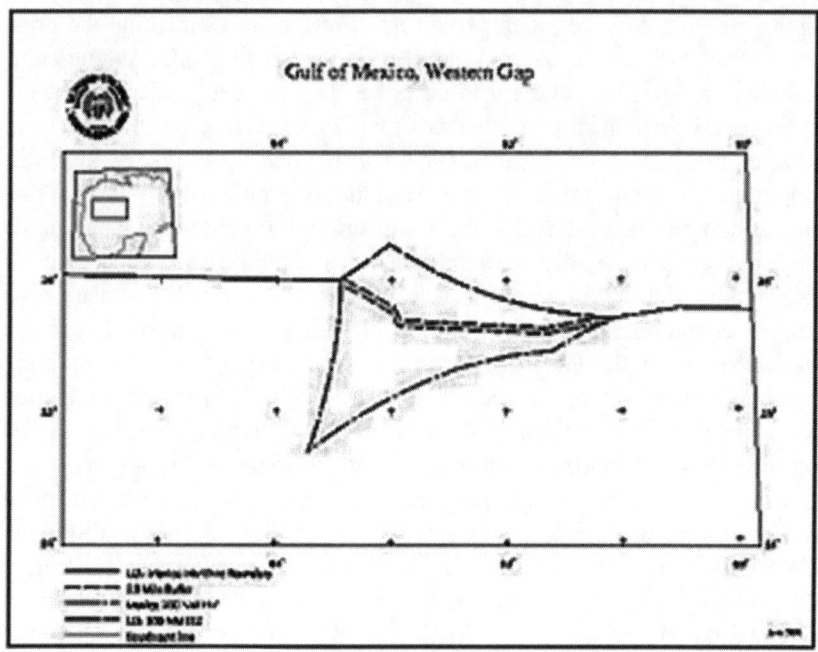

Image 3.1: Gulf of Mexico Western Gap

Routes to policy convergence (II): 1982 UNCLOS

Although the Reagan administration refused to sign the initial UNCLOS
treaty, the United States signed the revised version (and its annex) during
the Clinton administration. The relevant UNCLOS provision concerning
the Western Gap delimitation negotiations was Article 137 of Part XI,
which provides that '[n]o state shall claim or exercise sovereignty or
sovereign rights over any part of the Area or its resources, nor shall any
State or natural or juridical person appropriate any part thereof'. UNCLOS
provides that the 'Area' in question – the Western Gap – was under the

---

180 *Id.*, at 460.

exclusive jurisdiction of the International Seabed Authority.[181] In other words, UNCLOS prohibited the area from becoming subject to the jurisdictional rights of any independent country because it was beyond the limits of national jurisdiction. The ultimate question which faced the United States and Mexico as they delved into their delimitation negotiations was whether the Western Gap did fall under that jurisdiction, or whether it was to be considered part of the 'high seas'.

A strict legal reading of UNCLOS, such as the one initially undertaken by Mexico, suggests that the Western Gap is part of the International Area, and neither the United States nor any of its corporations would have authority to 'explore or exploit' the natural resources located within it. In contrast, the United States simply did not accept Part XI of UNCLOS. They adhered to the view that the Western Gap was a submarine area subject to the same principles that applied to the 'high seas' and was not under the jurisdiction of the International Authority. Therefore, it was 'open and freely available for use by all states, regardless of their location'.[182]

In order to resolve the UNCLOS issue on delimiting the Western Gap, special attention was drawn to Article 76(1) UNCLOS, a pivotal clause which classifies the continental submarine shelf into two different categories: the 'physical' shelf and the 'legal' shelf:

Article76

Definition of the continental shelf

1. The continental shelf of a coastal State comprises the seabed and subsoil of the submarine areas that extend beyond its territorial sea throughout the natural prolongation of its land territory to the outer edge of the continental margin, or to a distance of 200 nautical miles from the baselines from which the breadth of the territorial sea is measured where the outer edge of the continental margin does not extend up to that distance.

Under this article, Mexico would gain the legal right to extend its maritime jurisdictional boundaries to meet the natural boundary of the conti-

---

181 *See* UNCLOS Part XI Article 1.
182 *See* ARND BERNAERTS, BERNAESTS' GUIDE TO THE 1982 UNITED NATIONS CONVENTION ON THE LAW OF THE SEA, 36, 42 (1988), discussing the concept of 'high seas' under Article 87 UNCLOS.

nental margin, as well as 'sovereign rights for the purpose of exploring and exploiting the [natural] resources' located in this area.[183]

As it continued delimitation negotiations of the Western Gap with the United States, Mexico appeared to change its policy with respect to its interpretation of UNCLOS in light of its extended continental shelf. To be able to engage in diplomatic negotiations, Mexico had the burden of producing evidence that the 'natural prolongation' of its land territory to the outer edge of the continental shelf actually did exceed 200 nautical miles.[184] So, it did.

Routes to policy convergence (III): NAFTA

Robert Moran claimed that the factors connected to the economic and business significance of the North American Free Trade Agreement (NAFTA) across North America had to be understood from each individual country's perspective. One of the challenges was thus to understand NAFTA's 'threefold' dynamics and to draw upon this understanding in considering how Canadians, Americans, and Mexicans worked together over the long term as partners for their mutual benefit. The worldwide trend of increasing globalisation shaped and stood as the defining reason for the emergence of NAFTA. The shifts in the balance of power among these three countries and their corporations was felt in Canada, the United States and Mexico.[185]

Arguments for and against NAFTA found currency across North America as the cross-cultural dynamics of trade between two industrial nations and a developing one became substantial. As the dynamic of this trade agreement encompassed countries which were by their very nature of vastly different economic strengths, the debate about who the winners and losers were quickly spiralled towards stalemate. Many factors came into play as the three countries strove to come together within the free trade agreement. Between 1994 and 1999, there was a shift of operations from the US to Canada or Mexico which resulted in some US workers losing

---

183 *See* Vargas, *supra* note 1, at 473.
184 *Id.*, at 473.
185 ROBERT T. MORAN, UNITING NORTH AMERICA BUSINESS–NAFTA: BEST PRACTICES, 11 (2002.).

their jobs. At the same time, NAFTA was creating jobs in Mexico, where in some cities, employers even complained of labour shortages.[186]

In Mexico, Vicente Fox was Mexico's first president in seven decades to be elected outside of the *Partido Revolucionario Institucional* (PRI). Mexico was highly dependent on the United States, whose market accounted for approximately ninety percent of its exports. This US demand stimulated an industrial transformation along the border, with the increase in the '*maquiladora*' industry (the use of imported raw material to make goods and then re-export them, often to the United States) aiding in the increase of economic growth. However, the success of free trade and a modern industrial economy caused strains in Mexico's infrastructure along the border. NAFTA was to be just one example of globalisation and, as in most aspects of this worldwide trend, there were those who benefitted and those who felt its negative effects.[187]

As a result of NAFTA, Mexico was positioned to begin servicing the large and growing population of Hispanics in the United States. This niche market had the potential to become a substantial window of opportunity for smaller and medium-sized companies. By building strong relationships with the Hispanic communities who could ultimately form the backbone of their business, these small and medium-sized Mexican companies were establishing firm foundations for their future growth. Mexico was also focusing on its key competitive advantages – industrial commodities and low-tech consumer goods. While multinationals were expanding their operations in Mexico to include more complex processes in response to continuous downward pressure on costs, less technology intensive businesses allowed Mexicans to benefit quickly in a reciprocal way. The effects of NAFTA on Mexico and its US critique can be summarised as follows:

– The Mexican economy was tiny compared to that of the US, and NAFTA impacts were similarly small on the US economy – 117,000 Americans signed up for NAFTA displacement benefits. This com-

---

186 *See* NAFTA AS A MODEL OF DEVELOPMENT: THE BENEFITS AND COSTS OF MERGING HIGH AND LOW WAGE AREAS 103 (Richard S. Belous and Jonathan Lemco eds., 1995.).

187 In particular, *See* Isaac Cohen, *The NAFTA's Winners and Losers: A Focus on Investment, in* NAFTA AS A MODEL OF DEVELOPMENT: THE BENEFITS AND COSTS OF MERGING HIGH AND LOW WAGE AREAS 37 (Richard S. Belous and Jonathan Lemco eds., 1995.).

pared to 1.5 million jobs lost from closing factories, falling demand, corporate restructuring and the 2.8 million yearly job displacement. Moreover, US FDI in Mexico averaged less than USD 3 billion per year (from 1994); this amounts to less than 0.5 percent of US firms' total spending on plant and equipment.

– Other trends had a greater impact on US–Mexico trade, such as the peso crisis which made Mexican goods cheaper and US goods more expensive.

– NAFTA was not the significant impetus for free trade is was promoted to be, as the US already had low tariffs and Mexico had been in the process of liberalisation since the mid-1980s.

– NAFTA could not be wholly credited for helping Mexico emerge from the peso crisis because had Mexico restricted imports, it would have effectively hurt its own export industry (integration with the US via the *maquiladora* programme entailed US imports to export industrial commodities).[188]

Finally, as with any complex multinational trade agreement, NAFTA was affected by the business challenges which result from cultural differences. Within the widely diverse region that is North America were three nations, Canada, the United States and Mexico, each with distinct national cultures enriched by innumerable subcultures. This played a negative role on integration. Detractors asserted that there was insufficient consensus among the different groups of society about where NAFTA fit into the many other changes they felt taking place around them. Consequently, among the substantial avenues for improvement and growth, the most challenging remained the creation of a common environment with sufficient cultural sensitivity and skill effectively to bridge the threefold dynamics of NAFTA and promote cultural synergy through the cooperation and coordination of two industrial nations and a developing one.[189]

---

188 *See* Moran, *supra* note 14, at 18.
189 *See* Harley Shaiken, *The NAFTA, a Social Charter, and Economic Growth, in* NAFTA AS A MODEL OF DEVELOPMENT: THE BENEFITS AND COSTS OF MERGING HIGH AND LOW WAGE AREAS 27 (Richard S. Belous and Jonathan Lemco eds., 1995); *also* JONATHAN GRAUBART, LEGALIZING TRANSNATIONAL ACTIVISM: THE STRUGGLE TO GAIN SOCIAL CHANGE FROM NAFTA'S CITIZEN PETITIONS (2008).

Routes to policy convergence (IV): deepwater technology and oil reserves

In the United States the Minerals Management Service (MMS) of the Department of the Interior pushed for the treaty in support of exploration and development of the portion of the Gulf. By resolving the boundary dispute, the MMS could offer leases to private companies within the Western Gap without risking Mexican protests, which could delay development even further.

Nevertheless, it should be noted that both the oil industry and the US government were probably aware that PEMEX (*Petróleos Mexicanos* – the Mexican state petroleum monopoly or NOC) did not have the technology or the funding to develop the Western Gap. It does not seem unreasonable that the United States was at least partially motivated by a desire to 'get there first'.[190]

The other side of the coin was Mexico's urgency to settle the dispute as a defensive tactic. The location of Mexico's crude oil and natural gas production was predominantly in the two gulf states of Tabasco (27%) and Campeche (67%) in the South of the Gulf of Mexico (*see* Map 3.2); both states produced onshore and offshore, but the latter in shallow waters. In addition, Mexico had witnessed deepwater drilling around its boundaries by United States and Brazilian companies, and feared it would be left out of the game and loosing part of its potential reserves as PEMEX failed to invest its revenues in technological development.[191] However, most importantly perhaps, if Mexico did not act, it could endanger both PEMEX's *de facto* monopoly in the sea zone and affect the 'gravitational' sense of identity and sovereignty forged by oil property in Mexico since the 1917/1938 'Mexican Revolution'.[192]

---

190 *US–Mexico gulf treaty pressures rising*, OIL & GAS J., May 12, 1997, at 34.

191 See EDGAR W. BUTLER, JAMES B. PICK and W. JAMES HETTRICK, MEXICO AND MEXICO CITY IN THE WORLD ECONOMY (2001) 185.

192 According to one of the outstanding historians of the period, the Mexican Revolution stressed tendencies towards nationalism, Indianism and agrarian reform, with extraordinary importance being laid on the labour problem. What marked the Mexican Revolution was intervention of the State in economic production as new form of control, of prevision and balance, and the formation of a government with the participation of all social classes in a functional, democratic system or, in other words, organisation of all the sectors of production. In the words of Lázaro Cárdenas: 'formation of our own economy will free us from certain kind of capitalism, the incentive behind which is none other than obtain raw materials

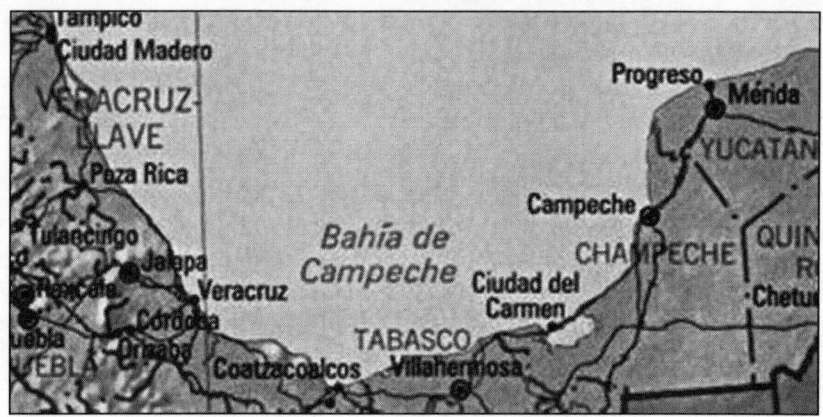

Map 3.2: Bahía de Campeche (Southern Gulf of Mexico)

Establishment of the maritime boundary: negotiating the TMDCS

To start negotiations, the governments of both the United States and Mexico eventually agreed on the idea that the equidistance method would probably yield the most beneficial results as the two parties negotiated the delimitation of the maritime boundaries in the Western Gap. This was the method ultimately used when negotiating the TMDCS.[193] Because the boundary lines were drawn equidistant from each country, in compliance with the International Law of the Sea, Mexico actually ended up receiving sixty-two percent of the Western Gap area (*see* Map 3.1, *ut supra*).[194]

The United States finally ratified the TMB and negotiations for the Western Gap began in late 1998. The negotiations were conducted by a Mexican delegation, headed by the Legal Advisor of the Secretariat of the Foreign Relations, and included representatives from the Secretariats of

---

by means of cheap labor; a capitalism that doesn't even decide to reinvest its profits in Mexico, that becomes a menace to nationalism in its troublous times, and that only leaves behind, in the last event, the exhausted lands, depleted subsoils, starvation wages, and unrest foreboding disturbance of order'. *See* Joe C. Ashby, *Labor and the Theory of the Mexican Revolution under Lázaro Cárdenas*, 20 THE AMERICAS (1963), 158-199.

193 *See* TMDCS Article 6.
194 *See* Gerald Karey, *U.S. Mexican Officials Agree Doughnut Hole Boundary*, PLATT'S OILGRAM NEWS, June 12, 2000, at 3.

Energy and of the Navy, the National Institute for Statistics, Geography and Cybernetics (INEGI), and from *Petróleos Mexicanos*. The US delegation was headed by the Legal Advisor of the Department of State and included officials from the Department of State and the Department of the Interior. US Deputy Assistant Secretary for Oceans and Space, Mary Beth West, led the negotiations with Mexico.[195]

Despite the fact that the use of the equidistance method for the delimitation of the Western Gap would give Mexico ownership of a majority of the mineral reserves in the area, one of the Mexican government's primary concerns for the delimitation negotiations was the protection of Mexico's sovereignty. Mexico was concerned about the drilling activities of private US drilling companies, such as Shell, Amoco, Mobil and Texaco.

Mexican politicians argued that jurisdiction over the maritime boundary should be promptly established. The Mexican opposition and media, on the other hand, believed that any reserves on the Mexican side simply belonged to Mexico, even if the reserve from which the minerals were extracted extended into the boundary area.[196]

There was also division in the US. On 29 June 1999, the Oklahoma-based coalition Save Domestic Oil, Inc. (SDO) filed a petition with the US Commerce Department. SDO alleged that Saudi Arabia, Mexico, Venezuela and Iraq dumped crude oil into the US market at less than fair market prices when the prices were down. However, the American Petroleum Institute (API) and the Interstate Natural Gas Association of America urged the Commerce Department to dismiss the petition, claiming that it lacked sufficient support from the US oil and gas industry. This was the path that the Commerce Department took, and SDO filed an appeal in September, 1999. Before SDO filed its petition, Mexico had imposed a four percent tariff on natural gas imports from the United States. Mexico had planned to lift this four percent tariff in July, 1999, which would have been ahead of the schedule outlined by NAFTA. SDO's petition, though, caused Mexico to reconsider its decision to lift the tariff, as political tensions rose as a result of the antidumping petition.[197]

---

195 *U.S., Mexico Soon to Delineate Gulf 'Gap'*, OIL & GAS JOURNAL, Dec. 15, 1997, at 27–28.

196 Ronald Buchanan, *Gulf 'Donut Hole' Focus of Study by PEMEX*, PLATT'S OIL-GRAM NEWS, May 22, 1998, at 3.

197 *See* Mathew Robinson and Eric Kronenwetter, *Dumping Motion Fifed: Mexico Eyes Ratification*, OIL DAILY, June 30, 1999.

In early November 1999 a negotiation meeting was held in Mexico City. Its two most important tasks were the establishment of boundary lines within the Western Gap and the determination of how natural resources would be addressed in the new treaty – the latter aim being regarded as the more politically sensitive of the two. Various principles were suggested to govern the treaty: (i) following the geography of the coastline; (ii) using an equidistant line as a basis for demarcation of clear borders; (iii) weighing alternative methods for revenue and responsibility sharing of industrial and environmental management; (iv) consulting technical experts and conducting pre-negotiation surveys; (v) studying previous treaties between the two countries to avoid contradictions; (vi) designing compliance measures; and (vii) maintaining equity. Only the principle of equidistance in delimitation prevailed.[198]

The design of the TMDCS

The delivery of the TMDCS was remarkable, as the process for determining territorial borders naturally marks 'the end of long and delicate negotiations, and [involving] an intricate balancing of legal, technical, and political considerations'.[199] As the result of successful negotiations, the United States and Mexico signed the TMDCS on 9 June 2000 in Washington DC. Secretary of State Madeleine Albright signed the treaty for the United States and Secretary of Foreign Relations Rosario Green signed the treaty for Mexico. On 18 October 2000 the US Senate consented to ratification of the TMDCS, and the Mexican Senate gave its approval on 28 November 2000. The United States and Mexico exchanged instruments of ratification in Mexico City at 11 a.m. on Wednesday, 17 January 2001.[200]

The TMDCS's *mandate* contains both distributive substance (Article 3) and process mechanisms (Articles 4 and 5). It is narrower than originally intended, as it focuses exclusively on oil and gas natural reserves without considering anything else. In this respect the TMDCS was innovative. To

---

198 *See* Jonathan I. Charney, *The American Society of International Law Maritime Boundary Project*, 5 MARITIME BOUNDARIES 10 (Gerald H. Blake ed., 1994).
199 *See* Vargas, *supra* note 5, at 190.
200 *See U.S. Mexico Complete Gulf of Mexico Treaty Ratification,* PLATT'S OIL-GRAM NEWS, January 19, 2001, at 6.

address Mexico's main concern, the exploitation of transboundary reserves, Articles 4 and 5 created a plan by which the parties have to transmit and exchange valuable information concerning the potential existence of reservoirs. Article 4(1) created what is known as a buffer zone, otherwise known as the 'Area', within which neither party may drill or produce petroleum for a ten-year period following ratification.[201] Once this ten-year period expires, both countries have the option of making the Area available for oil and gas exploration. Article 4(5) requires that each party share with the other any geological data in its possession concerning the existence of any transboundary reserves in the Area. If either country obtains any geological or geophysical data indicating a transboundary reservoir, the treaty requires both parties to seek and reach agreement for the 'efficient and equitable exploitation of such transboundary reservoirs'.

Dispute resolution is addressed in Articles 6 and 8, with respect to interpretation and application. As can be seen, the *rules* are legalistic and respect public international law both in their peaceful methods and cooperative *ideology*.[202] To conclude the description, it remains to mention that the treaty's *scope* is unintrusive: even though it makes reference to 'national laws and regulations' in Article 4(5), it does not impose any reform on the parties.

Cooperation: outcomes

After the entry into force of the TMDCS, once sovereignty was secured, things appeared to return to normal: that is, each country continued with

---

201 This buffer zone comprises the area extending 1.4 nautical miles outwards on each side of the 135-nautical mile boundary line. The United States and Mexico statistically determined that a 1.4 nautical mile buffer zone would provide a greater than 99.9% chance that neither party would be drilling or producing any petroleum that could be contained in a transboundary reserve, this respecting any natural resources that lawfully belonging to the opposing nation in the Area.

202 The Preamble states that the governments proceed '[d]esiring to establish, in accordance with international law, the continental shelf boundary' and '[c]onsidering that the practice of good neighborliness has strengthened the friendly and cooperative relations'.

its own development policy.[203] No joint research plans were made, setting aside in a sense the *norms* substance of the treaty.

In the United States during the treaty negotiations, the MMS had deferred offering 336 blocks (1,759,159 acres) in the Western Gap. All of these blocks lay in deep water, mostly at depths greater than 2,500 metres. MMS received bids on some blocks in the area during a 1997 lease sale; however, those bids were returned unopened. Therefore, the leasing programme remained within its original frontiers. Finally, no technical research was conducted locally, as the OCS Scientific Committee Deepwater Report Conclusions shows in its 2001 recommendations:[204]

> Recommendation: That in addition to identifying data sources under the study titled "Gulf of Mexico Deepwater Information Resources Data Search and Literature Synthesis", MMS make reasonable efforts to also obtain the data sets within budgetary and time constraints. Obtaining the data sets coincident with identifying them is time and cost effective as opposed to trying to obtain them several years from now.

> FY 1999–2001 Deepwater Research Plans: The subcommittee finds that, in general, the research proposed for FY 1999–2001 will provide MMS with the information required to make informed decisions on existing and anticipated deepwater lease and post-lease activities over the next 5–10 years. The FY 2001 studies plan is in the early stages of development and, based on recommendations from the subcommittee and others, additional studies will be incorporated as information needs become better defined. The studies strike an adequate balance between the disciplines and are properly timed and phased to provide information as it is required by MMS. The subcommittee, in its review, came up with several recommendations which it believes will strengthen MMS's deepwater research program. Additional funding may be necessary to fully implement these recommendations without impacting MMS's ability to meet its environmental information needs in other OCS areas.

> Initiate efforts to collect and compile existing information for that portion of the OCS bordering on the western gap; very little is known of this area of the GOM, and these data will be important when the U.S and Mexico eventually ratify a boundary treaty. When appropriate, initiate field efforts to collect additional data which are needed. A review of the existing seismic data now

---

203 For example, *see* U.S. Report of the National Energy Policy Development Group (2001). *Also* Crude oil Market Outlook 2008–2017, Secretaría de Energía, Ministerio de Economía de México (2008).

204 Available at http://search.icrard.org/cgi-bin/MsmGo.exe?grab_id=42&EXTRA_ ARG=&host_id=42&page_id=9118208&query=western+gap&hiword=western+ gap+.

held by the GOM regional office would be valuable in determining the complexity of this area that has been so heavily leased.

[…]

Recommendation: MMS should develop studies to further understand and evaluate the environmental consequences of FPSOs.

Gulf of Mexico Gap Areas: There are two areas in the Gulf of Mexico, one in the western Gulf and one in the eastern Gulf, which do not fall under the EEZ jurisdiction of either the United States or Mexico. The two countries are actively negotiating boundaries for the western gap area. This area is of particular importance to MMS and the offshore oil and gas industry since several existing OCS leases about the current US boundary of the gap. Thus, it is important that environmental and ecological information be collected for the western gap and adjoining areas and used to manage oil and gas activities which will take place once the boundary is agreed to by the two nations.

Finally, it should be noted that since 11 September 2001, the US has been more focused on planning for possible security breaches in the Gulf of Mexico than oil development.[205]

The same situation prevailed in Mexico, as no measures were taken. A Report was filed by the Department of Energy – *Secretaría de Energía* (SENER) – on 27 March 2008. A private company had requested 'a copy of any report, analysis or information that had been made [PEMEX] on development viability in the "doughnut holes"'.[206] The answer was that 'there was no document produced by [PEMEX] studying the possibility of oil extraction'.[207] After this, the Access to Information Institute – *Instituto de Acceso a la Información* (IFAI) – imposed on SENER the duty to ensure the means exist to provide any information requested.[208]

---

205 *See National Strategy for Maritime Security*, DEPARTMENT OF DEFENSE AND HOMELAND SECURITY (2005).

206 'Por medio de la presente, solicito de la forma más respetuosa a su dependencia copia de cualquier reporte, informe o análisis que haya realizado la paraestatal sobre la viabilidad de extraer petróleo de los llamados "hoyos de dona"' *Secretaría de Energía del Ministerio de Economía, Comité de Información*, RES: CI-IV-2008-019, at 1.

207 'A la fecha, no se cuenta con un documento realizado por la paraestatal Petróleos Mexicanos, en virtud del cual se haya realizado algún estudio, cuyo objetivo sea extraer petróleo de los polígonos occidental y oriental (Hoyos de Dona). *Id.*, at 2.

208 *Deberán informar de hoyos de dona*, EXELCIOR, 10 July 2008. (*They will have to provide information about the doughnut holes*, EXELCIOR JOURNAL, July 10, 2008.).

No substantial exploration is foreseen, despite efforts to change this through by some international organisations such as the International Committee on Regulatory Research and Development (ICRARD), which both the MMS and the Mexican Institute of Petroleum (IMP) are members of.[209]

Conclusions

The study-case enables the separation of the crafting of regimes into two stages, meaning the conclusion of the TMB was followed by the TMDCS.

Also, I conclude that even though there was a strong functionalist element to the Mexican sense of *identity* underpinning its desire to proceed, the true force which mobilised negotiations at both stages was the issue of *mistrust* in the other party's development of the offshore oil reserves. Evidently, functionalism imposed itself in the preliminary periods, in the treaty design, and in the later conduct of the states. Both the Mexican and the American executives had to deal with inner pressures, either from

---

209 ICRARD focuses on transferring knowledge in the areas of health, safety and the environment in the petroleum sector. ICRARD's purpose is to coordinate research activities, to exchange information and to promote research cooperation between these organisations. The ICRARD seeks to establish 'Terms of Reference' to provide a forum to advise on research and development activities funded by offshore regulatory authorities; exchange details of current research and development programmes on a regular basis; make available reports from completed research and development programmes to other authorities; co-sponsor research and development projects, when appropriate; and exchange information on research and development programme strategy. Most recently, ICRARD participated in a discussion on hurricane Katrina research with the Mexican Institute of Petroleum (IMP) and the Society of Earthquake Engineering (SMIS) during the 9th Symposium on Strategic Construction and Natural Hazards in Ixtapan de la Sal, Mexico, on 21 to 24 February 2007. ICRARD thus provided the forum within which the negotiations between the MMS and the *Secretaria de Energia* of Mexico concluded a Memorandum of Understanding (MOU) to enable the Parties to share scientific and technical information related to offshore oil, gas and mineral activities including but not limited to information on risk perception, safety of personnel and offshore installations, environmental protection, pollution prevention, pipelines, and Floating Production Storage and Offloading facilities, accident/incident reporting and the development and evaluation of regulations. Most importantly, this MOU would establish a procedure for cooperation. However, and again, negotiations remain at an early stage.

other state organs (congress), political parties or private sector actors. However, it was probably Mexico which had the hardest deal, as it had to modify both its internal normative and foreign policies. Ultimately, however, even though no material gains have yet been achieved, both countries achieved their goals: the United States got the Gap opened to future development and Mexico defended its oil sovereignty.

Commentaries are set out in overview in the Final Table below.

Treaty	Historical Context	Level of Equity *de facto*	Effects on Exploration	Effects on Production	Effects on Revenues	Regional (Social) Transcendence	Level of Cooperation
Western Gap	Pacific	Normal	None	None	None	None	Low

Table 3: Final Table

6 Africa: a long night's trip into the light

The scramble for reserve replacement has become increasingly intense in the first decade of the twenty-first century. The causes are threefold: a relentless increase in global oil demand coupled with a failure of the industry to develop new reserves to replace those being consumed and to meet the growing demand, fuelled in particular by China's rapid economic growth and the United States' new conception of oil supply as an object of national security policy.[210]

One of the more ominous explanations for the abovementioned failure of supply to keep pace with demand might be that global oil production has peaked. As a consequence, future discoveries will be inadequate even if demand were eventually to level off. A different explanation advanced by some oil companies is that restrictions on access to new resources imposed in some of the most promising prospective regions in the world have prevented these areas from realising their full productive potential. These factors have forced the oil industry to take maximum advantage of technological advances that enable exploration and production of oil and gas in remote, frontier areas, both on land and at sea, including offshore drilling in deep water. In Africa the situation could be the reverse, as otherwise apparently accessible areas are cut off for a lack of available knowhow. However, this access is not always 'transparent'. This chapter considers this situation through the dyad formed by Nigeria and São Tomé e Príncipe. The main goal is to discover if a joint agreement between the countries could impact on the inner and outer limitations of the factors affecting the accessibility of this region's oil reserves and these two nations' potential for development.

---

210 *See* Michael Klare and Daniel Volman, *The African 'Oil Rush' and US National Security*, 27 Third World Quarterly (2006) 609–628; *also* Michael Klare and Daniel Volman, *America, China & the Scramble for Africa's Oil*, 33 Review of African Political Economy (2006), 297–309.

Nigeria

Nigeria is the world's eighth largest oil producer and has the seventh largest reserves of natural gas. As a major oil exporter, Nigeria and the West African Gulf region are expected to supply a quarter of the United States' oil requirements by the end of this decade.[211] The country has already been in the oil market for many years, with onshore and offshore development:[212] it has become a member of OPEC[213] and has established a national oil company.[214] As noted by the Financial Times, 'the region's oilfields have become an important battle ground of influence between China, India, and the US as they struggle to ensure the motor of their future economic growth does not run out of fuel'.[215]

Among African countries, Nigeria has great potential to harness the opportunities and meet the challenges that the global economy can provide. Along with South Africa, they account for fifty-five percent of the region's GDP and a quarter of its total population. Their dominance is even more pronounced at the sub-regional levels. Within its immediate sub-region, Nigeria is one of fifteen countries in West Africa which make up the Economic Community of West African States (ECOWAS) and accounts for half of its population and three-quarters of its total GDP.[216]

Nigeria is a well-endowed country in terms of natural and population resources. The country is on course to continue the medium-term growth of 6.3 percent achieved during 2005–2007. Intending to reduce the incidence of poverty by a third over the long-term, an average annual GDP growth rate of 7.0 percent is projected by the government for the period 2005–2015. However, to realise its potential within the global economy,

---

211 *See NIGERIA – The Petroleum Reserves*, APS REVIEW GAS MARKET TRENDS, Monday, July 30, 2007, *available at* http://www.allbusiness.com/agric ulture-forestry-fishing-hunting/support-activities/4493935-1.html.

212 *See Nigeria – History of Oil Exploration*, APS REVIEW GAS MARKET TRENDS, Monday, August 3 2009, *available at* http://www.allbusiness.com/mini ng-extraction/oil-gas-exploration-extraction-oil-oil/12634580-1.html.

213 *See NIGERIA – The OPEC Decision Makers–Part 9*, APS DIPLOMAT OPERA-TIONS OIL DIPLOMACY, Monday, September 18, 2000, *available at* http://ww w.allbusiness.com/government/714853-1.html.

214 *See Nigeria Decision Makers–The NNPC Structure*, APS Review Downstream Trends, Monday, August 20, 2007, *available at* http://www.allbusiness.com/trade -development/international-trade-exports-imports-by/5495884-1.html.

215 *FT Africa Oil and Gas*, Financial Times, March 1, 2006.

216 MAX SIOLLUN. OIL, POLITICS AND VIOLENCE (2009.).

Nigeria needs to overcome several key challenges relating to governance, infrastructure bottlenecks and human development.[217]

Challenges to development (I): national politics

In the short to medium-term, Nigeria confronts two main challenges on the governance front. First, since the return to democracy in 1999, outbreaks of ethno-religious violence have gained momentum. It is estimated that 50,000 people have been killed in various incidents of ethnic, religious and communal clashes, with close to one million people internally displaced. Fighting has escalated in the Niger Delta, leading to hostage-taking and disrupting oil production. With the new administration in May 2007, the Nigerian government started to make concerted efforts to resolve these issues. Second, Nigeria is associated with a high level of corruption.[218]

Challenges to development (II): economic policy

Nigeria has so far been a minor player in the global economy, and in terms of global competitiveness, Nigeria is also undermined by a weak financial sector. With GDP at USD 100 billion, the country accounted for 0.28 percent of world's GDP in 2005. Another striking feature of Nigeria's economy is that it remains largely informal, with a significant share of transactions in the economy remaining unreported. At fifty-eight percent of GNP, the country's informal economy was estimated to be one of the ten highest in the world. Nigeria's score was twice the average of 28.8 percent, with China having the lowest score at 13.1 percent. As a ratio of GNP, the informal economy in Nigeria was six times higher than that of the United States (8.8%). While the informal nature of the country's economy may be an indicator of an inspired national entrepreneurial spirit, it reflects a low level of development in the global economic context. It also has implications on the design and delivery of formal fiscal, monetary and structural policies as well as measuring the impacts of these policies. The challenge

---

217 REGIONAL ATLAS ON WEST AFRICA (Lurent Bossard ed., 2009.).
218 SOALA ARIWERIOKUMA. THE POLITICAL ECONOMY OF OIL AND GAS IN AFRICA: THE CASE OF NIGERIA (2009.).

has always been to bring the wide range of informal sector activities into the formal sector without discouraging the innovation and dynamism of these activities.[219]

When compared to other African countries, the private sector in Nigeria is relatively well-established and diversified. The private sector consists of informal micro and small-scale enterprises as well as medium and large scale corporate businesses, often with joint ownership between Nigerians and foreigners. Private sector businesses are diversified across sectors, which include agriculture, industry, financial services, trade and commerce. Historically, the high cost of doing business in Nigeria, due to factors including corruption, administrative barriers and poor infrastructure, have hampered the development of the private sector. The picture which emerges is that business competitiveness in Nigeria could be further improved by streamlining the administrative requirements and obstacles to the private sector.[220]

Running alongside this, Nigeria lacks a modern and efficient physical infrastructure to underpin and sustain productivity, competitiveness and global integration. The poor state and inadequacies of Nigeria's systems of roads, rail, ports, airports, telecommunications and power generation and distribution contribute to the high costs of doing business there. In particular, generation and delivery of electricity is unreliable, with the private sector incurring huge costs for standby power generators. Small firms in particular, devote around a quarter of their capital budgets for self-generation of electricity. The electricity sector suffered considerable neglect throughout the decade of 1990s. No new power stations were built between 1990 and 1999, no major overhaul of existing plants was carried out, and only nineteen out of seventy-nine generating plants were in operation in 1999. As a consequence, on average, firms provide their own electricity sixty-seven percent of the time, at a cost of more than twice that provided by the national electricity agency. The nationally installed power

---

219 Bosworth, Barry, P. and Susan M. Collins. *The Empirics of Growth: An Update*, *in* BROOKINGS PAPERS ON ECONOMY ACTIVITY (1996) 113–179. Also *Nigeria Private Sector Assessment*, THE WORLD BANK REGIONAL PROGRAM ON ENTERPRISE DEVELOPMENT, *available at* http://www.usaid.gov /ng/downloads/reforms/nigeriaprivatesectorassessmenttechnicalpapers.pdf.
220 *Id.*

generation capacity was 4,200 MW, while peak power generation languishes at a maximum of 3,500 MW.[221]

Because of this, Nigeria has been eager to make economic reforms, thinking that they might pay off, even if slowly, in terms of improving the business climate. Since 1999, the Government has embarked upon a comprehensive programme to divest its interest in state-owned enterprises (SOEs), which has cut across thirteen key sectors, ranging from agriculture to aviation and from the oil and gas industry to the hospitality industry. The cumulative value of investment to be transferred from the public sector was estimated at about USD 100 billion. The Bureau of Public Enterprises noted that within a relatively short period, over twenty-five mostly smaller-public enterprises in various sectors of the economy had been privatised, including in the banking industry, insurance, cement and sugar manufacturing, oil and gas, hospitality, shipping, vehicle assembly and media. The returns on investment on these enterprises has been very encouraging.[222]

However, progress has been slow in the sale of the larger SOEs, such as the port authority, the aluminium smelter company, the power authority and the telecommunications company (NITEL). NITEL is Nigeria's national fixed line and mobile telecommunications company. It was fully owned by the Federal Government of Nigeria (FGN). NITEL, in turn, owned 100 percent of M-Tel, the analogue mobile cellular communications company. NITEL and M-Tel's operations were merged in 2001. NITEL had a dominant position in fixed line telecommunication with 720,000 lines and approximately ninety percent of the market. This provides potential investors with a strong customer base and an existing network which could be expanded.[223]

Finally, though Nigeria could have invested into its oil and industry infrastructure, it remained a net exporter of capital.[224] It is well-known that capital flight from Nigeria and sub-Saharan Africa (most of which is

---

221 *Id.*
222 *See Nigeria Private Sector Assessment*, THE WORLD BANK REGIONAL PROGRAM ON ENTERPRISE DEVELOPMENT, *available at* http://www.usai d.gov/ng/downloads/reforms/nigeriaprivatesectorassessmenttechnicalpapers.pdf.
223 *Id.*
224 *See* for example, Neil Ford, *Where does all the oil money go?*, AFRICAN BUSSINESS, Thursday, June 1, 2006. *Available at* http://www.allbusiness.com/a ccounting/accounts-receivable/4097569-1.html.

unreported) is huge. The Commission for Africa noted that around forty percent of African savings are held outside the continent, compared with just six percent for East Asia and three percent for South Asia. Despite the scarcity of capital for productive purposes, Africa slightly exceeded even the Middle East (39%) in the high proportion of private wealth held abroad. It is estimated that capital flight from sub-Saharan Africa was about USD 15 billion per year, roughly equal to the aid flows into the continent.[225]

Challenges to development (III): demography and the professional diaspora

In terms of population, Nigeria has the third largest number of people living in poverty in the world, after China and India. Around seventy million Nigerians, fifty-two percent of the population, live on less than USD 1 a day (the UN Human Development Index 2005 ranked Nigeria 158). As a consequence of poverty, the human development indicators in Nigeria are extremely poor: in 2003, net primary school attendance stood at sixty percent (64% for boys versus 57% for girls). About one million Nigerian children under the age of five died (an infant mortality rate of ten percent of live births, one of the highest in Africa). Life expectancy at birth is estimated at 43.4 years. The prevalence of HIV/AIDS is estimated at five percent, compared to an average of 1.2 percent for all developing countries and 7.7 percent for Sub-Saharan countries.

Nonetheless, Nigeria had a large pool of potential consumers and workers projected to increase by almost two-and-a-half-fold from 2000 to 2040. As Nigeria's population is relatively young and fast-growing (2.4 percent), it is projected to reach 356 million by 2050, placing the country as the fourth largest in the world behind India, China, and the United States, respectively. Nigeria's demographic trends will impact long-term economic development through factors including labour supply and productivity, savings and investment behaviour, government fiscal activities, pensions requirements and social welfare provision. The population's youthfulness has major implications for employment, education and the provision of other government services. As the demand for basic social and

---

225 *See* ARIWERIOKUMA, *supra* note 3.

infrastructure services rises, social and economic pressures on the nation's limited resources are likely to increase. However, Nigeria's large pool of potential consumers had not yet been fully tapped, because per capita income was and remains low. In 1999 Nigeria's per capita income stood at USD 300, which made it half that of India, at USD 715, even before adjustment for purchasing power parity (PPP). Adjusted, India's per capita income was three times that of Nigeria.[226]

In this context, it is a necessary though harsh conclusion that for Nigeria to be competitive in the global economy, development of its human resource is as essential as its infrastructure. Leaving aside the abovementioned percentages, the quality of skills development in Nigeria was affected by poor facilities for science and technical education. To address the challenge of poor human and social development, from 2006, debt service costs due to Paris Club nations to the tune of USD 1 billion were redirected to the social sectors of education and health.[227]

Another economic link to the rest of the world involves the movement of people, particularly skilled professionals. Skilled workers from Nigeria represent two-thirds of the total emigration from the country to OECD countries, slightly higher than the skilled emigration rates for India (60.5%) and South Africa (62.6%). This brain drain's downside is exacerbated by the fact that the education of skilled emigrants was funded through public subsidies. However, emigration has also had positive economic effects through return migration, owing to the additional knowledge and skills migrants acquire abroad, the creation of trade and business networks and remittances from abroad.[228]

In a sense, Nigeria's future economic prospects and relative position depend on participating in an increasingly global knowledge economy. Nigeria is behind on measures to stimulate the development of the knowledge economy, including measures related to literacy, skills, research and development, patents, and information technology. With respect to intel-

---

226 *See* Beth Anne Wilson and Geoffrey N. Keim, *India and the Global Economy*, 41 Business Economics (2006) 28–36.
227 *See* ARIWERIOKUMA, *supra* note 3. *Also Doing Business in 2006: Creating Jobs*, THE WORLD BANK AND INTERNATIONAL FINANCE (2006), available at http://www.doingbusiness.org/Downloads/.
228 For Nigeria, remittances now average USD 1.5 billion per year, including around USD 1.3 billion per year from the United States alone (around 1.9% of GDP in 2004). For India, remittances are estimated to amount to between three and four percent of GDP. *See* Beth Anne Wilson and Geoffrey N. Keim, *supra* note 12.

lectual property, the World Trade Organization noted that Nigeria recorded no patents or registered marks in 1999.[229]

São Tomé e Príncipe (STP)

A former Portuguese colony, STP gained its independence in 1975. In 1990 the country adopted a new constitution providing for multi-party democracy. STP is a tropical island nation. Located in the Gulf of Guinea off the West African coast, the country consists of two main islands and a number of smaller islets, which together have a total land area of just over 1,000 square kilometres. It is the smallest country in Africa, with a population of less than 200,000. The country imports all of its fuel, most manufactured and consumer goods and a significant portion of its food. Its productive base is undiversified as it relies almost exclusively on cocoa exports and external donations.[230]

From cocoa monoculture, through an unviable state, to a petro-state

In fact, cocoa is not STP's only export commodity: its other important agricultural exports include coffee and copra. STP has some potential for exporting non-cocoa foodstuffs, notable among which is its Monte Cafe coffee, which is highly prized by international connoisseurs. However, successive governments have failed to diversify agricultural production. Foreign-assisted projects from the reintroduction of cattle by the Dutch to the establishment of a poultry farm by the Cubans were entirely unsuccessful. Therefore, the value of non-cocoa exports remains small and cocoa exports continued to account for almost seventy-five percent of the country's total exports of goods in 1999.[231]

Until the 1980s, the centre of economic activity was the country's plantations, but by the 1990s, the centre had shifted towards the nation's capital, São Tomé. This has had profound implications for the relationship between town and country and the elites and society at large. Within São

---

229 WORLD TRADE ORGANIZATION (2005), *available at* www.stat.wto.org/Cou ntryProfiles.
230 *Supra* note 4.
231 *Id.*

Tomé city, the most economically active sectors are commerce and government. These sectors continue to be largely driven by foreign aid and loans rather than export earnings. Commercial trade is mostly in imported goods, while the islands produce little more than a selection of basic foodstuffs and beverages. Such production had little potential for generating dollar revenues and, unsurprisingly, domestic agricultural production is of little interest to the decision-makers in São Tomé, whose livelihoods depend on external sources of income.[232]

While STP failed to build on its own resources, successive governments have been very adept at attracting aid from a range of donors for the last twenty years, exploiting both the superpower rivalries and historic ties with Portugal (which was the biggest provider of bilateral aid in the 1990s). During the Cold War, the Soviet Bloc (notably East Germany) represented a major provider of assistance, although help was also garnered from various Western countries and the People's Republic of China. The latter was supplanted by Taiwan in 1997, with aid inflows of some USD 10 million per year (which makes Taiwan the biggest source of bilateral aid today) apparently being conditional on ongoing Santomean diplomatic recognition of that country. As a result of external assistance, foreign aid inflows constituted some twenty to twenty-five percent of GDP by 2000. STP has received millions of dollars from multilateral institutions such as the UN and the World Bank. Whilst STP was still firmly part of the Soviet Bloc, the Movement for the Liberation of São Tomé and Príncipe (MLSTP) government signed an agreement with the World Bank and the IMF on the rehabilitation of the cocoa plantations through foreign management financed by multilateral credits in January 1984. Three years later, the first Structural Adjustment Programme was launched. Further structural adjustment policies in the 1990s allowed for more access to aid funds. The STP government often failed to control public spending and to implement agreed economic reforms, which led to temporary suspension of World Bank/IMF programmes in 1990 and in 2001. However, successive STP governments were ultimately able to continue receiving substantial aid from the World Bank and the IMF and other multilateral institutions. In addition, with its vast external debt (standing at some USD 300 million or over USD 2000 per Sao Tomean in early 2002) the country qualified for assistance under the Heavily Indebted Poor Countries (HIPC)

---

232 *Id.*

initiative, in return for further reform. As a result of STP's dependence on foreign aid and loans, STP became what is termed 'an unviable state' even before the end of the Cold War and has remained unviable ever since.[233]

The country owes its continued existence to an international system which guarantees the survival of small internationally recognised, albeit unviable, state entities. State sovereignty enabled the microstate to obtain considerable external resources, whether from the World Bank in exchange for promises of policy initiatives, or from Taiwan in exchange for extending diplomatic recognition. The trappings of sovereignty offered much greater potential for generating hard currency than cocoa exports ever could.[234]

While STP has done rather well out of attracting foreign aid and loans, the 'interregnum of unviability' seems likely to be supplanted soon by another major transformation – into a 'petro-state'. Oil production was expected to being within four to ten years in 2003 and could provide the Santomean government with access to far greater financial resources than ever before.[235]

Oil exploration had already been undertaken in STP during the colonial era. In 1970 the colonial authorities granted a Hydrocarbon Exploration PLC, a subsidiary of a British firm Ball and Collins, and the US firm Texas Pacific Oil Company. Seismic studies indicated the presence of hydrocarbons and Hydrocarbons drilled two wells in 1973, although the company later withdrew because of the high cost of drilling. In the 1980s exploration was undertaken by Island Oil Exploration. The firm conducted seismic studies and drilling onshore and by 1990 was said to have invested

---

233 In 1996 external development aid to STP totaled some USD 38 million, while STP's exports of goods totaled USD 4.9 million. To continue speaking of a cocoa economy in STP would be a total misunderstanding of the country's position in the international political economy. In order to demonstrate STP's transformation towards an unviable state, one only needs to compare the country's export revenue with its imports. According to IMF estimates for 1999, STP exported goods worth some USD 3.9 million (of which cocoa accounted for USD 2.9 million), while it imported goods worth USD 21.9 million (of which food alone made up USD 4.8 million). STP's agricultural exports did not even suffice to cover the country's own food needs, let alone generate a surplus. *See Regional Economic Outlook: Sub-Saharan Africa*, IMF (2006); also, *World Economic Outlook*, IMF (2006.).

234 *Supra* note 4.

235 *Id.*

over USD 2 million in exploration in STP. However, the emergence of STP as a potential petro-state did not occur until the late 1990s.[236]

Problem (I): national politics

International interest in STP's potential as an oil producer in the late 1990s can be attributed to technological advances which allowed for the possibility of oil production in the ultra-deep offshore areas in the Gulf of Guinea. From an oil industry perspective, STP's territorial waters have great geological potential. Most notably, the geological structures of the Niger Delta extend into STP's territorial waters in the north. Oil exploration in the territorial waters of STP's neighbours was highly successful in the late 1990s, with prominent oil discoveries such as Agbami, Nnwa and Akpo on the Nigerian side to the north and Ceiba on the Equatorial Guinean side to the east.[237]

Even before the oil majors became interested in STP, the STP government had entered into various memoranda of agreement, letters and stipulations regarding the evaluation and study of oil and gas reserves with the Environmental Remediation Holding Corporation (ERHC), a small US firm, and at certain times with the South African firm Procura Financial Consultants (PFC). The government led by Prime Minister Raul Neto signed an agreement in June 1997 with ERHC and PFC (as junior partner) to negotiate on behalf of the country with other oil companies for the award of all government concessions in the sector. In July 1998 the government and ERHC established a joint venture petroleum company, Sociedade Nacional de Petroleos de São Tomé e Principe (STPetro), with the government holding fifty-one percent of the shares. One month after the creation of STPetro, the US oil major Mobil and STPetro signed a letter of intent to enter into a technical assistance agreement to evaluate the early hydrocarbon potential of the offshore seabed located in the country's exclusive economic zone (EEZ). Under the terms of the exclusive agreement, Mobil completed an eighteen-month technical evaluation in STP oil blocks numbered 1 through 22, which had previously been scheduled for an open oil-licensing round. Geco-Prakla (now Western-Geco) – a

---

236 *Id.*
237 *Id.*

Schlumberger subsidiary which had previously been hired by ERHC as a technical advisor – conducted seismic studies. Following successful exploration, Mobil selected five of the twenty-two oil blocks earmarked for exploitation. This could make the now merged ExxonMobil one of STP's key oil producing firms in the future, although the company's right to the oil blocks is not legally watertight and the terms were renegotiated in January 2003.[238]

STP's emergence as a potential 'petro-state' highlighted the country's lack of human resources to deal with the new tasks of state diplomacy. As a consequence, the STP government has had to rely on external actors to deliver its petroleum policy. ERHC therefore initially usurped some of the country's sovereign rights as the firm was solely responsible for the development of STP's entire oil and gas sector. ERHC apparently even arranged meetings with UN and US representatives for President Miguel Trovoada and Prime Minister Raul Neto during their state visits to the US. In other words, a private firm replaced an arm of the STP diplomatic body. In addition, the government remained dependent on the assistance of neighbouring states. Traditionally, STP has been assisted by Angolans. Former STP prime minister and president Miguel Trovoada explained that Angola's national oil corporation Sonangol helped the STP government in negotiations with Mobil in 1998–1999.[239]

On the other hand, as in every oil-dependent country, conflicts between members of the elite obviously broke out, as demonstrated by the deteriorated relations between the new President and Miguel Trovoada after the 2001 presidential election. In September 2001 Menezes dissolved parliament, a move which was contested by most political parties and caused anxiety about political instability and delays in oil licensing in oil industry circles. On 4 February 2002, Patrice Trovoada, Trovoada's son and STP's foreign minister, was even forced to resign, after accusations by Menezes that the Trovoada family were treating the country as their private fiefdom. While ideological issues were usually of little importance, access to foreign aid money and other public funds caused or exacerbated many, if not most, intra-elite conflicts. Many conflicts between President Trovoada and the various prime ministers in the 1990s were attributed to rifts over foreign aid or business contracts. The extraordinarily close interconnec-

---

238 A POLITICAL AND ECONOMIC DICTIONARY OF AFRICA (David Sedon *et al.* eds., 2005.).

239 *See* ARIWERIOKUMA, *supra* note 3, at 101.

tions within the elite facilitated corruption and the lack of accountability, which is discussed in greater detail above. Compared with other African countries, the spoils of foreign capital inflows could be shared within a smaller elite. While political events have seen power shifting out of the hands of the Trovoada family – who were in power at the time the various oil deals were made – neither the new president nor the successive prime ministers have demonstrated a serious desire to prosecute those involved in the ERHC/Chrome or PGS deals, rhetoric notwithstanding. The durability of Menezes's 'technocratic' new order depend on his ongoing ability to sideline established political actors and movements. This reluctance reflects the nature of elite formation in São Tomé, which centres on shifting alliances between the key political leaders and their factions. Periodic, seemingly bitter quarrels were resolved with new deals being made.[240]

A positive remark must be made before continuing with this description. Unlike mainland tropical Africa's recurrent political strife, marked by large-scale ethnic and regional divisions, STP's small population means that there are relatively few people who need to be persuaded to accept any shifts in the balance of power (or paid off).[241] Nonetheless, despite the very peaceful political environment, STP has experienced two military coups – in 1995 and 2003 – but these were mostly bloodless events and were quickly resolved peacefully, ending with the restoration of civilian rule in both cases. An oil boom is thus unlikely to trigger the violent conflicts it has elsewhere in Africa, yet it may breed further corruption and may introduce further economic distortions. The Santomean political elite have proved remarkably adept at shoring up the fortunes of what turned out to be an unviable state, profiting from Cold War rivalries, colonial era links and sentimentalities, and the sale of the trappings of statehood, ranging from flags of convenience to the diplomatic recognition of Taiwan. However, whilst it would seem likely that they would have materially benefited from a number of recent petroleum deals, their inexperience in this

---

240 *See. Regional Economic Outlook: Sub-Saharan Africa,* IMF (2006). Also, *World Economic Outlook,* IMF (2006). In addition, A. Kraay Kaufmann and M. Mastruzzi, *Governance Matters IV: Governance Indicators for 1996–2004,* THE WORLD BANK (2005), *available at* www.worldbank.org/wbi/governance/pubs/govnatters4.html.

241 *Id.*

area suggests that individuals based on the African mainland will gain a great deal more.[242]

Problem (II): international politics

In the international arena, concerns about STP's petroleum deals were expressed by the World Bank and the IMF. These two institutions concluded that oil revenues from previous oil exploration agreements were managed with little transparency by the STP government. The IMF stated in its usual diplomatic language that 'oil sector negotiations in 2001 had lacked transparency and the ERHC settlement diverted STP potential oil revenue, raising serious governance problems'.[243]

It is important to mention that unlike elsewhere in Africa, democratic reform did not start with the end of the Cold War, as concrete steps had already been set in motion in the late 1980s, with the invitation of exiled senior politicians to return to the country and the introduction of some political reforms. In December 1989 the STP parliament approved adopting liberal democracy in a new constitution, though the elite party – the MLSTP – still hoped to continue ruling in this changed political landscape. However, the democratisation process developed its own dynamics, culminating in the first democratic multi-party elections in 1991 for both the presidency and legislature, which saw the MLSTP voted out of power and supplanted by the Democratic Convergence Party (or *Partido de Convergência Democrática* – PCD-GR) in the legislature. Miguel Trovoada, then standing as an independent with the blessing of the PCD-GR (but now of the Independent Democratic Action Party or *Acção Democrática Independente* – ADI), won the presidential poll. Whilst STP has experienced frequent changes in parliamentary majorities since 1991, the new system has proved durable, with hotly contested elections held regularly. However, while STP has all the appearances of a stable democracy, in sharp contrast to other parts of Africa, the political landscape remains dominated by a handful of families and factions, and is characterised by shifting alliances and temporary marriages of convenience. Seemingly bitter conflicts are regularly resolved in a manner which is bewildering to the

---

242 *Id.*
243 *Id.*

outsider. The system permits the accommodation of competing interests in such a manner that existing relations or patterns of authority are never in jeopardy. The willingness of the losers to accept election results and wait for the next time, and for ruling political parties to tolerate serious rivals stands in sharp contrast to mainland African states such as Zambia, Zimbabwe or Namibia. However, this is not simply due to the pragmatic cohabitation of political parties and to an extraordinary ability to compromise on day-to-day policy but also due to strong social bonds between the members of the political elite. The differences between the main parties in Santomean politics are often only superficial. Personalities and personal connections matter most. Indeed, the key players in STP politics have all had long-standing social bonds. All of the new major political parties after 1990 – the PCD-GR, the ADI and the MDFM – emerged out of factions led by former senior MLSTP officials. Trovoada was previously MLSTP prime minister in the 1970s and Menezes was a top cabinet minister in the 1980s. When Trovoada stepped down from the presidency in 2001, Menezes was his anointed successor, having previously served as a parliamentarian for Trovoada's party, ADI.[244]

It is of course possible that some adjustment to the process of privatising state resources can be made. Many foreign observers believe that President Fradique de Menezes' criticism of previous oil sector contracts was genuine and, even if it will not lead to any prosecutions, it may lead the STP government renegotiating them on more favourable (if still comparatively modest) terms. Such a turn of events was backed by the World Bank and the IMF, which have become increasingly irritated with the Santomean failures to implement economic reforms and the propensity for the elite to conclude self-serving deals. Indeed, in May 2002 President Menezes announced his intention to abrogate or renegotiate the previous oil contracts and agreements, including the contract with Chrome Energy, the STP-Nigerian JDZ and ExxonMobil's oil licenses. Menezes had previously requested technical assistance from the World Bank to scrutinise the contract with Chrome Energy and to conduct a cost-benefit analysis of it, although this was reportedly a precondition set by the IMF staff. Little is known in public about this. However, it has been reported that Rafael Branco's two children were among those who had received college scholarships from an ERHC-monitored programme in STP. The World Bank

---

244   *See* Seldon *et al.* eds., *supra* note 24.

funded a confidential review of existing contracts and treaties for the STP government, which was conducted by a Houston-based law firm. Menezes also turned personally to a multitude of US actors for legal and technical advice and assistance in renegotiating oil contracts. These included the likes of the Washington-based consultancy Petroleum Finance (a specialist energy consultancy), Joe Kennedy and his company Citizens Energy, Yale University and the law firm Williams & Connolly (which previously represented the former US president Bill Clinton).[245] In the end, privatisation was mainly promoted by the African Oil Policy Initiative Group (AOPIG), whose members include representatives from ChevronTexaco (which produces vast amounts of oil in Angola and Nigeria, among other places), the minor US oil firm Vanco Energy (which controls vast oil concessions in Africa, including in Equatorial Guinea, Morocco and Namibia) and US State and Defense Department officials.

President Menezes has also strongly supported the creation of a naval base, not least because this would provide STP with a deepwater port, which the country lacks (and which the Nigerians failed to deliver despite some promises made in the past). However, in October 2002, the US Assistant Secretary of State for African Affairs, Walter Kansteiner, denied reports that the US was planning to build a military base in STP, though he did reveal that the US was still exploring other ways of expanding military cooperation with STP, including by providing patrol boats.[246]

The Nigerian government was greatly displeased by this course of events. President Obasanjo flew to São Tomé in early June 2002 insisting on the sanctity of previous oil contracts and the Nigerian government made serious threats against STP. However, Nigerian pressure did not alter the position of Menezes, who could count on the protection of the US government, which had pledged military assistance to STP, beefed up security at its Voice of America relay station in STP and had – at some point – even considered the establishment of the aforementioned regional naval base for aircraft carriers and patrol boats in STP. In a departure from the strategies of his predecessors, who relied on Angolan, Gabonese or Nigerian patronage, it would appear that Menezes is hoping to be able to rely on US protection.[247]

---

245 *Id.*
246 *Id.*
247 *Id.*

Indeed, despite all the presidential rhetoric about the unfair oil contracts, the Santomean government has been able to reach an accommodation with the foreign oil companies. In early 2003, the government renegotiated its oil related contracts, including those with ExxonMobil and ERHC/Chrome. While some of the most onerous terms have been eliminated from the new ERHC/Chrome contract – i.e. terms which entitled the oil company to a share of future government taxation – ERHC/Chrome was able to gain concessions elsewhere. As a result, the Nigerian owners of ERHC/Chrome have been left with a highly favourable contract. Meanwhile, President Menezes admitted that he previously received a secret payment of USD 100,000 from ERHC's chairman, which he claimed was a political party contribution. While this payment appears to have been made before the cooling of relations with Nigeria and seems to have been unrelated to the contract renegotiation in early 2003, it tainted the President's reputation. It also confirmed our prediction, made in September 2003, that Menezes' rhetoric about renegotiating oil contracts would amount to little more than minor changes in contract terms.

Of course, like the past experience of other oil-rich states, corruption and mismanagement are likely to be exacerbated by the arrival of oil money and the affairs of ordinary Santomeans are likely to become even further removed from the minds of the political decision-makers. Perhaps the best hope for STP is that there is sufficient external pressure from international institutions, creditors and the incipient civil society to ensure greater transparency in the distribution of oil revenues. Otherwise, STP is likely to suffer many of the same ills as other oil-rich states in Africa, though any civil war or social unrest is highly unlikely in the gentle Santomean society. STP has always been very peaceful and, from this perspective, a highly positive role-model for the continent. Unlike other emerging or existing African petro-states, STP did not experience violent internal conflict (like Sudan or Chad), it did not have a history of mass murder and internal repression (like Equatorial Guinea), and it did not witness any ethnic or religious clashes (like Nigeria).[248]

---

248 *Id.*

Nigeria–São Tomé e Príncipe Joint Development Arrangement

In addition to the internal difficulties experienced by Nigeria and STP already described, a major obstacle for both countries was the undetermined maritime boundaries with their neighbouring countries. Many times, uncertainty over property rights have deterred major oil companies from committing themselves to investments in their territorial waters. Indeed, the exclusive deal in STP Mobil was able to clinch (which apparently includes the right to first refusal on all blocks) was because other major companies such as Exxon and Shell are believed to have backed away from the offering owing to this lack of clarity.

To resolve these obstacles, the countries signed the 'Treaty between the Federal Republic of Nigeria and the Democratic Republic of São Tomé e Principe on the Joint Development of Petroleum and other Resources' on February 21, 2001, to establish legal premises of an international nature to support the development of their sea based resources without further negotiations regarding sovereignty.[249] The solution they found was to establish a Joint Development Zone (JDZ).

The JDZ area lies between latitudes 1° and 3° north and longitudes 4° and 8° east in the Gulf of Guinea (Map 4.1). The JDZ is estimated to cover an area of 34,548 km^2 in water depths of 1,500 metres in the northern sector up to 4,000 metres in the south-west. The area lies proximate to the Golden Rectangle, which has reserves of about 5 billion barrels of oil.

---

249 Article 4.

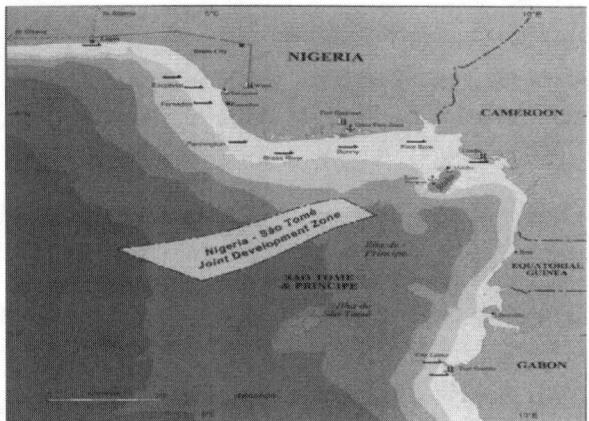

Map 4.1: Gulf of Guinea JDZ

Source: Nigeria – STP Joint Development Authority: Guide to the 2004 JDZ Licensing Round Document.pdf

As has been noted several times, where international boundaries are entangled in economic and political problems, companies are only willing to undertake oil and gas related activities in these areas if both governments make a commitment to establish a business-friendly environment. Nigeria and STP opted for joint development and the Treaty concluded by the Presidents of the two countries has this principle ingrained in it basic provisions, which include:

- A Joint Ministerial Council;[250]
- A Joint Development Authority;[251]
- A petroleum regime for the JDZ;[252]
- Regulatory norms (tax, criminal provisions, health, safety and the environment);[253]
- Disputes resolution.[254]

---

250 Part Two (Articles 6 to 8).
251 Part Three (Articles 9 to 13).
252 Part Eight (Articles 21 to 31).
253 Part Ten (Articles 36 to 46).
254 Part Eleven (Articles 47 to 49).

Joint Ministerial Council

The Joint Ministerial Council which under the Treaty was established to perform specific functions necessary to advance and achieve many of the Treaty's basic objectives. It provides that the Council at all times comprises of a minimum of two ministers or a maximum of four ministers or persons of equivalent rank appointed by the Head of State of each party. The Council is also empowered to deliberate on matters concerning the activities of the Joint Development Authority (JDA) and is required in addition to perform the following functions:

- give directives to the Authority on the discharge of its functions;
- approve rules, regulations and procedures for the effective function of the Authority;
- consider and approve audited accounts and audit reports of the Authority;
- consider and approve the annual report of the Authority;
- review the operations of the Treaty and make recommendations to the State parties on any matter concerning the functioning or amendment of the Treaty, as appropriate;
- approve a development contract which the Authority proposes to enter into with any contractor;
- approve the termination of development contracts entered into between the Authority and contractors;
- consider and approve the annual budget and the opening of bank accounts by the Authority;
- settle disputes in the Authority; and
- approve the appointment and remuneration of auditors.

Joint Development Authority (JDA)

In establishing the JDA the Treaty provides that it must report to the Council, have a judicial personality in international law and enjoy similar status and capacity under the laws of each of the States party. Such capacity is to enable it to perform its legitimate duties. In practical terms it is the administrative organ established to administer the activities of the JDZ. It is accountable to the Council and has several functions, which include:

- demarcate the JDZ into Blocks, prepare bid documents and execute the bid process;
- execute development contracts with contractors subject to Council approval;
- supervise the activities of contractors;
- recommend to the Council the termination of the activities of contractors;
- allocate to the States party the financial or other benefits in proportion to their participatory interest;
- control the movement of vessels, aircraft, equipment and structures into, within and out of the JDZ;
- regulate and direct on all matters related to the supervision and control of operations in addition to health, safety and environmental issues;
- regulate marine activities in the JDZ;
- preserve of the marine environment, bearing in mind the relevant provisions and international laws governing the JDZ;
- collaborate with other agencies in the generation, collation and exchange of scientific, technical and other data concerning the JDZ and its resources;
- prepare annual reports for submission to Council;
- advise and recommend to states parties on issues concerning material changes to the law which may be necessary to promote the development of the resources in the JDZ.

Boundary dispute negotiation

International boundary disputes arise as a result of claims and counter-claims by sovereign states over land, space or territory which lie within a contiguous geographical location. In this context therefore, the desire to negotiate is be required or guided by many factors including:

- the economic, cultural or occupational value attached to the disputed territory;
- an expansionist philosophy and the perceptions of an individual state;
- a perceived threat to national security;
- the commercial value of the disputed land; and
- questions of logistics and accessibility.

Boundary disputes are often very extensive in nature. Several methods can therefore be adopted to resolve them. Tumsah opined that boundary disputes can be resolved through the following:[255]

- Negotiation – this involves dialogue between the affected parties and may culminate in establishing a mutually accepted boundary line.
- Arbitration – this involves the use of third parties or experts who understand the issues and deep feelings involved in the dispute and are thus able to achieve a compromise acceptable to both parties.
- Litigation – states involved in boundary disputes often resort to litigation in order to have a judicial body examine the facts and deliver an unbiased judgment. The ultimate objective in this case is the demarcation of the disputed area.
- Wars – territorial disputes create tension and such tension, if not properly managed, can lead to war. The conflict between Iran and Iraq originated from territorial disputes over oilfields. The same is true of the dispute between Iraq and Kuwait. The conflict between Israel, Palestine and Lebanon can also be viewed as deriving from disputes linked with unresolved boundary delimitation issues. In some of the cases examined here for the reference countries, situations deteriorated into bloody wars and terrorist attacks.
- Establishment of JDZs – JDZs are usually a product of extensive negotiations between sovereign states which have overlapping boundaries. The purpose of the establishment of such a zone is to create an opportunity for both parties to benefit by jointly developing the affected area.

JDZ oil and gas regulations

It is important for oil and gas activities in a region to be carried out in an orderly manner. Accordingly, the Treaty provided for oil and gas regulations to guide concessionaires' conduct and activities. The JDA serves as the custodian of all areas within the JDZ as well as Oil Prospecting Licences (OPLs) and Oil Mining Licenses (OMLs) originating from the JDZ. In this regard the JDA is empowered to act as follows:

---

255 Tumsah, K.M., *Nigeria – DRSTP Joint Development Zone: A Legal Perspective*, 1 JDZ News (2005) 8.

- grant, subject to approval of Council, Exploration Licences (ELs), OPLs, OMLs and Production Sharing Contracts (PSCs) to companies operating in the JDZ registered in Nigeria and STP respectively;
- supervise all operations carried out under OPLs, OMLs and other contracts concluded by the Authority;
- enjoy at all times unlimited access to all areas covered by the OPLs, OMLs and contracts and other installations which are operated or maintained in support of these licences, leases and contracts to carry out inspections or related activities;
- summon in writing the holders of OPLs, OMLs, contractors or their subcontractors to appear before the Authority at a specified time and place for the purposes of providing information pertaining to its operations;
- order the suspension of operations carried out under the OPLs, OMLs and contracts in order to provide an environment which is safe and conducive to carrying out exploration and production or other related activities.

Partitioning the zone into Blocks

One of the functions of the JDA is the demarcation of the JDZ into distinct blocks of appropriate dimensions for the purpose of granting OPLs. Each block demarcated is for practical purposes assigned a distinct identifying reference number. The boundaries of the blocks are further defined if necessary by meridians of longitude of whole units of 5 minutes and in the same manner by parallels of latitude in units of 5 minutes.

Exploration License (EL)

An exploration license accords the licensee the right to undertake speculative geophysical surveys in a specified area. The holder of an EL may drill holes not exceeding 300 feet below the ground provided such drilling is for the purposes of facilitating geological and geophysical work. The grant of an exploration license in a particular area does not necessarily preclude further grant of license or lease in oil and gas exploration and production. An EL may not automatically convert into an OPL or OML and enjoys a life span specified in the agreement. In accepting the grant of an EL the licensee undertakes at all times to carry out its activities under close supervision of qualified personnel and such supervisory activities must be conducted to the satisfaction of the JDA.

Oil Prospecting Licence (OPL)

An OPL grants a licensee the right to explore for and exploit oil and gas within an assigned block. Upon discovery of oil in commercial quantities it may be converted into an OML. In the case of an OPL, the licensee is entitled to drill holes to any depth in the furtherance of exploration and production objectives. The licensee may also evacuate all or part of all discovered equity oil on condition of fulfilment of all the obligations placed on it by the terms of the OPL – royalties, rents, petroleum tax regulations and other applicable laws. The JDA provides that an OPL should enjoy an initial term of about four years, which upon formal application in writing may be extended for an additional mutually agreed period.

Oil Mining Lease (OML)

The OML confers on the lessee the exclusive right to 'search for, win, work, carry away and dispose of all petroleum in, under or throughout the Block described in the schedule'.[256] The Treaty provides for a term of twenty years for OMLs, with a renewal clause as contained in the PSC (see below). The nature of the lease in the Nigerian context is not quite clear, especially with respect to the fact that prior to 2006 no categorical decisions have been made in the courts concerning the nature of the lease. In 2006 SAPETRO applied to the government to convert portions of OPL 246 into additional OMLs but this was denied. Pursuant to this decision the federal government of Nigeria took steps to recover the unutilised portions of OPL 246 following the conversion of the OPL into an OML. The recovered portion of the block was later awarded to the Oil and Natural Gas Corporation (ONGC) of India. The lessee took steps in the Court of Appeal of Nigeria to restrain the federal government in its effort to repossess the unconverted portion of the block. In light of this the court ruled

---

256 M.M. OLISA, NIGERIAN PETROLEUM LAW AND PRACTICE (1997) 23.

that the Minister of Petroleum, guided by the Petroleum Act 1969, acted properly in not granting SAPETRO's request to convert the OPL into additional OMLs. Under these circumstances SAPETRO appealed the judgment to the Supreme Court. This was perhaps the first time such a case would be decided by a court of competent jurisdiction. In examining related issues,[257] Olisa expressed the view that the nature of the grant of an OML in the Nigerian context is imprecise as to whether the lessee enjoys a freehold, a tenancy at will, personal property or an interest in the land. It is important to note, however, that there is always substantial litigation concerning the juridical nature of a lessee's interest in petroleum matters in countries where oil and gas laws have developed over a long period.

Production Sharing Contract (PSCs)

The activities in the JDZ are executed through PSCs. Furthermore, the Authority is empowered by the Ministerial Council to enter into contractual agreements with companies incorporated in Nigeria or STP. With respect to PSCs, the Authority can exercise additional functions as follows:

- execute a PSC with a contractor with a view to granting the contractor exclusive rights to undertake petroleum activities in the assigned OPL;
- ensure that the contractor is obliged to subject its activities to the supervision of the Authority and observe all regulations in respect of the PSC;
- ensure that every PSC embodies specific undertakings by the contractor to the Authority to observe and perform all obligations set out in the contract;
- take appropriate steps against a licensee, lessee or contractor in relation to an event which pertains to terms of the contracts;
- place all the available blocks on offer for competitive bids;
- develop a model PSC for Ministerial Council approval and adopt the same as a standard PSC for exploration and production activities.

---

257 *Id.*

Norms	Scope	Mandate	Formal Rules	Agents Autonomy
Economic Liberalism	Intrusive (countries obliged to take internal measures)	Mixed	Highly elaborate and detailed	High autonomy

Table 4.1: Treaty Characteristics

Summing up

The Treaty designs a highly legalised institution (see Table 4.1). A broad and self-executing mandate is informed by detailed and elaborated rules. In addition, there is high delegation and autonomy in the bodies created to manage the JDZ. In this sense, it seems reasonable to affirm that the regime aims to solve the problem of corruption by creating a new environment for the achievement of high-tech offshore development. However, both countries admittedly remain obliged to take all the internal measures needed to make the Treaty truly operational.

Treaty implementation

The Gulf of Guinea remained unnoticed for many years but recent events in the region have brought it to prominence. Extensive seismic activities over several years have revealed the region to be oil bearing. Initial estimates indicate that the region has reserves ranging between five and fourteen billion barrels of oil. Some gas reserves have also been discovered. As pointed out, Nigeria and STP agreed in principle and further concluded a Treaty which – pending the final demarcation of the overlapping maritime boundaries of the respective states – will enable the overlapping zone to be developed jointly. A JDA was created to further this mutual understanding. For the purposes of allocation and contract execution, part of the JDZ was initially demarcated into nine blocks. These blocks (1–9) were auctioned on 23 August 2003. All licensing round tenders undertaken by the JDA are subject to competitive bidding by interested parties. The tendering is rigorous in nature and involves several processes which culminate in the award of blocks to successful candidates.

The available records indicate that twenty indigenous and multinational corporations submitted thirty-three bids. The number of bids notwithstanding, the Authority was constrained by the quality of the bids and the technical and commercial competence of the companies to award only one block to a consortium of ChevronTexaco (51%), Esso (40%) and Dangote Equity Energy Resources (DEER – 9%). In furtherance of the developmental objectives for the zone, blocks 2 to 6 were put on offer in 2003. The PSC for block 1 has been signed and negotiations on blocks 2 to 6 have been concluded, paving the way for the signing of the relevant PSCs.[258]

Oil Block bid process

The tendering process administered by the JDA as a first step invites tenders from interested parties. The information provided includes: the blocks earmarked for allocation, the bidding system to be adopted, details of the tender agreement to be executed, the rights and responsibilities of potential concessionaires, the period within which applications must be submitted.

Applications for competitive tenders

The intention to participate in the operations of any of the blocks on offer is registered with the JDA through a written application. Such applications would be filed with a set of documents or information as follows:
– proof of payment of prescribed processing fee;
– proof of the applicant's good financial standing and technical capability;
– details of the work programme to be executed on the block;
– the signature bonus offered by the applicant;
– detailed outline of the proposed annual expenditure;
– specific date when operations will start following the granting of a tender agreement by the JDA;

---

258 *Id.*

– detailed recruitment programme of staff from both party states and schedule of regular performance reporting.

The formation of the JDZ has strengthened the relationship between the two countries. The level of interest in the region continues to grow. It is expected that the areas of cooperation would increase, thereby paving the way for fisheries and other resources to be exploited in the region. A major consideration is the security provided by the region and the potential for a steady supply of oil and gas from the region. Nigeria and the US have already set in motion a Gulf of Guinea energy security initiative. This initiative will create an enabling environment which will promote the safe execution of oil related activities in the region.[259]

Political benefits of the JDZ

Prior to the formation of the JDZ by Nigeria and the STP there were no tangible bilateral activities between the two countries. The determination on the part of the collaborating States led to a mutual agreement to explore and exploit natural resources jointly in the overlapping maritime boundaries.

Formal demarcation of the maritime boundaries was deferred to a later date. The creation of the JDZ and the JDA has paved the way for officials of the Nigerian Boundary Commission and Ministry of Petroleum Resources to visit STP regularly to pursue the objectives of the JDA. Offices have been opened in Nigeria and STP to administer the activities of the Authority. The two nations have become closer to one another and areas of collaboration have extended to commerce. In July 2003 a military *coup d'etat* was staged in STP by disgruntled military officers of the island state. One of the demands was the abrogation of the Treaty between Nigeria and STP establishing the JDZ and the JDA. Sadly, the incident occurred during an official visit of President Fradique De Menezes to Nigeria. Despite the embarrassment to Nigeria of the ill-advised action of the military, President Obasanjo intervened and established dialogue with the recalcitrant officers. The issues at stake were resolved and the coup was reversed. As a mark of solidarity, President Obasanjo accompanied President Fradique De Menezes to STP to ensure that his return to the

---

259 *Id.*

island state was uninterrupted. Both countries have been cooperating in other mutually beneficial areas. The successful execution of the Treaty by both countries has demonstrated the capacity of countries in the region to amicably resolve boundary-related problems. Other political benefits derived from the conclusion of the Treaty between Nigeria and STP are as follows:[260]

- amicable resolution of boundary disputes devoid of prolonged litigations;
- strengthening of bilateral cooperation and peaceful coexistence in the sub-region;
- promotion of regional integration, especially in the context of the New Partnership for African Development (NEPAD) and the African Peer Review Mechanism (APRM);
- promotion of democratic culture;
- promotion of Extractive Industry Transparency Initiative (EITI);
- exchange of Nigerian industry technocrats to train personnel in STP.

Economic benefits of the JDZ

The JDA has commenced activities and a number of blocks in the JDZ have been awarded to successful bidders.[261] Consistent with the industry practice, companies were required to pay a signature deposit as a precondition for securing the blocks. In 2004 blocks 1–6 were opened for bids by interested companies. Block 1 was offered to ExxonMobil and Environmental Remediation Holding Corporation (ERHC). A total of USD 406 million was raised from the companies' signature deposits, and an additional USD 1.51 million was raised through the sale of bid forms and seismic data information. Other blocks will be auctioned in the near future and attract inflow of sustainable revenues into the treasuries of the collaborating states. Chevron has also announced plans to invest USD 20 billion in the African region. Part of the investment will be in the Gulf of Guinea. The revenues generated through signature deposit and other related activities will be applied to the execution of projects and programmes that will

---

260 *Id.*
261 *See NIGERIA – Foreign Partners' Role In Deep-Water Oil*, APS REVIEW OIL MARKET TRENDS, Monday, August 8, 2005. *Available at* http://www.allbusine ss.com/mining/oil-gas-extraction-crude-petroleum-natural/494833-1.html.

boost economic activities in the African sub-region. The exploration and production programmes will also spin-off other revenue generating activities in the oil services subsector, fabrication, financing and the provision of other professional services. The aggregate expenditure in the Gulf of Guinea has the potential to have profound positive economic impact in the region through the creation of a multiplier effect. There is an opportunity for other countries in the region with overlapping maritime boundaries to emulate the spirit of cooperation demonstrated by Nigeria and the Democratic Republic of São Tomé e Principe.[262]

São Tomé e Príncipe's individual outcomes

Authors diverge their opinions of the positive and negative consequences of the agreement.

As a positive political consequence, they mention that despite their differences, the small political elite at STP eventually found ways to cooperate. Following the inconclusive legislative elections in March 2002, which gave no one party an absolute majority in parliament, a power-sharing deal was concluded involving all the parties which had won representation in parliament in the March elections. In April 2002 a new coalition government was sworn in, headed by the supposedly politically-independent former Santomean ambassador to Portugal, Gabriel Costa. Costa had previously belonged to at least two different Santomean political parties. On taking office, he stressed that all of the main parties would have input into government policies, which would be aimed at redressing a range of social ills, from 'anarchy' and a lack of public service ethics, to robbery and vandalism.[263]

However, some national political commentators, in the early days of the 'transition' period, took a less general view and focused on certain very specific shortcomings of the impact of oil revenues on politics.[264] Even

---

262 *See* PRINCETON LYMAN. NIGERIA'S ECONOMIC PROSPECTS: AN INTERNATIONAL PERSPECTIVE (2004.).

263 *See* Seldon *et al.* eds., *supra* note 24.

264 For example, it is asserted that while the STP government had only officially received no more than USD 15–20 million from ERHC/Chrome, PGS and ExxonMobil combined by the end of 2002, the huge future potential of STP's oil has already had a pronounced impact on STP government policy. The STP government certainly lacked experience in dealing with oil company managers and

though oil production has not yet started in STP, petro-money has already started to influence Santomean politics. The 2001 and 2002 elections were perhaps a harbinger of things to come in that money seemed to play a greater role and Santomean political movements were assisted by Nigerian, Angolan, Gabonese and Taiwanese financial assistance. In an interview, former President Miguel Trovoada stated that, while the polls were very peaceful, the outcomes of the elections were marred by the intervention of money. This may be something of an exaggeration, but it is probably correct that Trovoada's party was outspent in the 2002 election campaign.[265]

Economically, STP looks set to become an oil producing state in few years. According to hypothetical future projections by the IMF, STP could produce almost 100,000 barrels of oil per day in less than a decade (although these figures were highly speculative). This would come close to the production levels of the other established producers in the region, such as Equatorial Guinea and Gabon. On the contrary, claims were later made that ultimately, a small island society was being impelled 'backwards into the big world' of unregulated markets, free trade and commodities movements, and was confronted with both the smiling face and seamy underbelly of the international petroleum industry. This would have far-reaching consequences for its elites – who, in their scramble for resources seemed already to have been short-changed – and the masses, who are likely to be confronted with many of the social ills of modernisation, whilst experiencing few of its benefits.[266]

---

lawyers, but much of the country's petroleum policy to date also lacked transparency and accountability. There does not seem to be any similar precedent in the history of Africa's oil industry since the end of colonialism. It is an unparalleled development to award so many oil blocks and prerogatives to a single company without a bidding process, let alone a small obscure firm with few notable assets which was on the verge of bankruptcy at one point. It is clear that political connections were key to the various agreements. *See* Seldon *et al.* eds., *supra* note 24.

265 *Id.*
266 *Id.*

Nigeria's individual outcomes

Nigeria's 'new path' has also elicited commentary. Associated with a high level of corruption, economic governance improved considerably between 2002 and 2007. The government embarked on various anti-corruption measures including the creation of the Economic and Financial Crimes Commission and the Independent Corrupt Practices Commission, as well as the removal of high-ranking civil servants from office on the grounds of corruption. To improve accountability, the government introduced the Fiscal Responsibility Bill and the Public Procurement Bill, while establishing a Budget Monitoring and Price Intelligence Unit in charge of public sector contracting and procurement contracting. To increase transparency in the dominant oil sector, the country took steps to participate in the G8 Extractive Industries and Transparency Initiative, under which full audits of the annual oil accounts of relevant agencies were to be implemented and published.[267]

As regards economics, since Nigeria's return to democracy in 1999, the Government has undertaken a wide range of economic and structural reforms, which have started to improve economic performance. It formulated the National Economic Empowerment and Development Strategy (NEEDS) covering the period 2003 to 2007 and made progress in implementing its economic reform programme. The objectives of NEEDS includes accelerating economic growth, reducing poverty, and meeting the Millennium Development Goals. NEEDS has three pillars: empowering people and improving social service delivery; promoting economic growth, particularly in the non-oil sector; and enhancing the effectiveness and efficiency of government and improving governance (Nigerian National Planning Commission, 2004). NEEDS has been designed to promote macroeconomic stability and to sustain higher economic growth and diversification. Structural reforms have also emphasised boosting external competitiveness through export-oriented reforms and prudent fiscal, exchange rate and monetary policies.[268]

In 2006 real GDP growth was estimated to be 6.2 percent, with non-oil GDP growth at about seven percent. During 2001–2005, real GDP growth averaged 5.6 percent per annum compared to 2.6 percent during the previ-

---

267 *See* Lyman, *supra* note 47. *Also* Seldon *et al.* eds., *supra* note 24.
268 *Id.*

ous five years. Higher economic growth benefited from the improved macroeconomic environment, higher oil export receipts and policy initiatives to spur agricultural production. The investment rate has averaged slightly more than a fifth of GDP in the past three years, while the savings rate is now at forty percent GDP. During this period, inflation rate averaged fifteen percent due to higher food prices and insufficient sterilisation of oil revenue inflows, but it was projected to decline to single digits in 2006. Fiscal deficits have been declining since 2003, reflecting a more prudent fiscal stance and increased oil export revenue. The government has been implementing a conservative, oil price-based fiscal policy, resulting in large overall budget surpluses projected at eighteen percent of GDP on a commitment basis and a significant build-up in international reserves, which reached USD 43 billion or 14 months of imports in early 2007.[269]

There is dichotomy between Nigeria's oil and non-oil producing sectors. The oil sector contributes around eighty percent of total government revenue, ninety-seven percent of total exports, and twenty-five percent of GDP. The country has abundant energy resources including an estimated 35 billion barrels of oil and an estimated 180 trillion cubic ft of natural gas, with reserves expected to last forty years and 110 years respectively, at current daily production rates. Nigeria aims to boost oil producing capacity from three million barrels per day to four million barrels per day by 2010 and five million barrels per day by 2020, with reserves projected to reach 40 billion barrels by 2010 and 50 billion barrels by 2020, respectively. More than half of its gas production is flared, with the remaining exported or used for electricity generation. It is estimated that gas flaring in Nigeria causes more greenhouse gases than the rest of sub-Saharan Africa combined. Despite this, Nigeria, which is Africa's largest oil producer, is already one of the biggest gas producers in the world and is projected to rise near the top of the list over the next decade. The production of liquefied natural gas (LNG) is projected to become critical to world energy needs as global demand for LNG is expected to rise by ten percent annually during the next decade. The United States and other Western countries are aiming to diversify their sources of hydrocarbon supplies away from the Middle East. Nigeria is capitalising on this trend with its

---

269 *2006 Budget Speech*, FEDERAL GOVERNMENT OF NIGERIA (2006), *available at* www.budgetoffice.gov.ng.

Nigeria Liquefied Natural Gas (NLNG) project, Africa's biggest capital project.[270]

There are also new trends on the oil industry. While major Western oil companies had historically dominated oil investment in the country, Asian oil companies have long begun to make inroads. For example, in 2006 China bought a forty-five percent stake in one of Nigeria's oilfields for USD 2.3 billion. Nigeria then provided priority rights on several oil blocks for state-owned Asian oil companies in exchange for infrastructure investment in power plants, refineries and railways. However, concerns were raised about the transparency of such large oil deals.[271]

Nevertheless, the pace of economic liberalisation and financial sector reforms in Nigeria has accelerated since 2004. Financial sector reforms increased bank capitalisation tenfold in two years. While Nigeria successfully concluded a debt forgiveness agreement with the Paris Club, the country has also been rated by leading credit rating agencies. In 2005 the IMF also approved a two-year Policy Support Instrument (PSI) for Nigeria under the IMF's newly created PSI framework, which was created to support the nation's economic reform efforts.[272] A satisfactory review of the benchmarks for the PSI paved the way for the clearance of the debt to the Paris Club in April 2006.[273]

However, since 2005, Nigeria's financial sector has been restructured, with the number of banks reduced from eighty-nine to twenty-five and the with minimum capital requirements increased tenfold. The financial sector reform process was widely acknowledged as one of the most far-reaching in the world. As a result of the reforms, Nigeria now has the fastest growing banking sector in Africa, attracting over USD 1.5 billion of foreign

---

270 *IMF Executive Board Approves a Two-Year Policy Support Instrument for Nigeria*, IMF Press Release, available at www.imf.org/external/pubs/ft/scr/2005/cr052 29.htm.

271 *See Nigeria – Part 2 – The Oil & Gas Fields And Foreign Operators*, APS REVIEW OIL MARKET TRENDS, Monday, August 10, 2009, *available at* http://www.allbusiness.com/africa/1011432-1.html; also, *NIGERIA – Foreign Partners' Role In Deep-Water Oil*, APS REVIEW OIL MARKET TRENDS, Monday, August 8, 2005, *available at* http://www.allbusiness.com/mining/oil-gas -extraction-crude-petroleum-natural/494833-1.html.

272 *See* LYMAN, *supra* note 48.

273 *Id.*

investment since 2005.[274] The financial sector, however, remains underdeveloped relative to the size of the economy.[275]

Nevertheless, despite these positive developments, key challenges lie ahead for Nigeria. As a nascent and fledgling democracy, it occupies a strategic position in Africa and is the key to stability in the region. Violence in the oil-rich Niger Delta has affected oil production and exports. Despite its oil wealth, Nigeria ranks among the twenty-five poorest countries in the world in terms of social indicators. Investments in social, human and physical infrastructure are also crucial to sustained long-term development.

Recent investments in the power sector are projected to increase generation capacity to 5,200 MW and 10,800 MW by end of 2006 and 2007 respectively (FGN, Budget Speech 2006). However, the transmission network remains inadequate, and it is estimated that about a third of the power generated is lost. As a result, only ten percent of rural households and about forty percent of the total population have access to electricity. Water supply and sanitation coverage are estimated at sixy percent and forty percent, respectively. The country has a road network of 195,200 km, with half in a poor state due to low maintenance.[276]

---

274 Before the reforms, no Nigerian bank featured among the global top 1000 banks. By 2006, twelve Nigerian banks were in the global top 1000. *See* Temitope W. Oshikoya, *Nigeria in the global economy: Nigeria's integration into the global economy is below its potential,* BUSSINESS ECONOMICS, Tuesday, January 1, 2008, *available at* http://www.allbusiness.com/economy-economic-indicators/output-demand-gross/11408248-1.html.

275 In 2006 credit to the private sector from the banking sector was 23 percent of GDP in Nigeria, compared to 14 percent in 2004, but still much lower than 80 percent in South Africa. Non-performing loans declined from a quarter of total loans in 2004 to seven percent in 2006, compared to five percent in South Africa. Only one in ten Nigerians had a formal bank account, compared to nearly five out of ten South Africans. The capitalisation of Nigeria's stock exchange is less than ten percent of the Johannesburg stock exchange in South Africa. Market capitalisation amounted to 20 percent of GDP compared to 214 percent in South Africa. The value of stocks traded on Nigeria's stock market amounted to 2.3 percent of GDP compared to 77 percent in South Africa, while turnover is three times more in South Africa. *See* Temitope W. Oshikoya, *Nigeria in the global economy: Nigeria's integration into the global economy is below its potential,* BUSSINESS ECONOMICS, Tuesday, January 1, 2008, *available at* http://www.allbusiness.co m/economy-economic-indicators/output-demand-gross/11408248-1.html.

276 *See* Temitope W. Oshikoya, *Nigeria in the global economy: Nigeria's integration into the global economy is below its potential,* BUSSINESS ECONOMICS,

Cooperation balance

As it was exposed at the beginning of the chapter and described in the previous pages, both Nigeria and STP have large oil reserves and low development capacity. These two characteristics resulted in a dyad of IOCs keen on offshore development, and corrupt governments looking to exploit it. On the other hand, recognition of the offshore development potential in their overlapping claim zone the and IOC financial pressure to improve their political establishment, imposed on both countries the decision to achieve an international development regime. This regime has proved profitable for both countries in many ways.

It seems clear that nothing but material interests drove actions initially. Nevertheless, I think it is fair to conclude that the promise of economic revenues persuaded political actors in both countries of the benefits of 'transparency', and encouraged initiatives to achieve greater transparency. Of course, these are mere tendencies and exceptions, as both administrations remain 'spoiled', and political actors and IOCs benefitting under the status quo are unlikely to wish to reform it.[277]

In any case, Nigeria and STP seem to have taken notice of oil market forecasting, and have taken some preliminary measures for joint action. Their conduct could be a model for other African countries in similar circumstances.

Tuesday, January 1, 2008, *available at* http://www.allbusiness.com/economy-eco nomic-indicators/output-demand-gross/11408248-1.html.

277 Neil Ford, *Where does all the oil money go?*, AFRICAN BUSSINESS, Thursday, June 1, 2006, *available at* http://www.allbusiness.com/accounting/accounts-recei vable/4097569-1.html; also, *Nigeria – Nigerian Politicians & IOC Attack New Petroleum Reforms Bill*, APS REVIEW OIL MARKET TRENDS, Monday, August 3, 2009, *available at* http://www.allbusiness.com/government/governmen t-bodies-offices-legislative/12634621-1.html.

7 Conclusions

To conclusion, I will summarise the studies performed, focusing on three issues. First, I will look at the variations in institutional design (the first dependent variable) among the cases as a function of variation in the matrix of independent variables identified in the introduction. Second, I will look at the variation in the nature of cooperation across institutions (the second dependent variable). Here I will also highlight similarities and differences in the efficacy across regimes. Finally, I will highlight some tentative findings about the relationship between regime design and the nature of cooperation.

Although a matrix for the research design was presented at the beginning of this work, the analyses have revealed that the variables were irrelevant in some cases. Nevertheless, it was possible to find some common ground and draw meaningful conclusions about similarities and differences.

Variations in institutional design and their sources

At the outset, it might be useful to recall that the literature on shared natural resources has been in most cases either descriptive or indicative, or purely based on the personal opinion of the author, without taking much notice of what goes on beyond the letter of the law. Though the original project intended to range more widely, four cases were examined above.

In the case of Middle Eastern countries, the agreements proved to be extremely short and narrowly focused on the point, that is, to dispose of existing or potential sovereignty problems to allow the contracting states to return to their ongoing hydrocarbon development activities. Rulers just wanted to consolidate their position in power through legitimacy, and keep themselves as distant as possible from regional conflicts, in order to deal with their European, American or Far Eastern clients. Pro-forma identity lines were introduced, but the general substance of the treaties was 'joined, but unconnected'.

In the Gulf of Mexico, the deal struck was different. Relations between Mexico and the United States had existed for a long timed and were moving towards a policy of 'bon voisingage',[278] expressed by the Rio Grande and NAFTA regimes. On the other hand, oil policies had long been clear by the time the Western Gap regime was agreed – national in Mexico and private in the US – and both executives did everything possible to secure sovereignty in the Gulf's remaining sea territory.

Finally Africa, where internal politics confronted international pressure, which ultimately prevailed and opened the door to unbelievable potential for growth.

Sources of institutional design

The main distinctions in the regime characteristics which influenced regime design are summarised in the table which follows.

---

278 *See* Alan K. Henrikson, *Facing Across Borders: The Diplomacy of Bon Voisi-nage*, 21 INTERNATIONAL POLITICAL SCIENCE REVIEW 121–147.

	Type of cooperation problem	Ideology and Identity	Systemic and sub-systemic power distributions	General international normative and economy scenery	Domestic politics	External institutions and non-state actors	Technical characteristics	History
Middle East	Functional problem relatively unimportant	Economic liberalism; discursive regional identity	Not important	Normative influence extremely important due development of sovereignty over natural resources and LOS regimes	Regime legitimacy and survival critical; similar regimes	IOCs important	Unimportant	Discursive, pro-forma importance
North America	Functional problem important due to mistrust between parties	Economic liberalism; regional identity desired	Not important	LOS regime important due to the need to overcome new provisions	Access to oil essential as the basis of legitimacy (Mexico) and 'machinery' functioning (US)	Unimportant	Important due to asymmetry in deep off-shore capacities	Important due to continuity in bilateral relations
Africa	Functional problem partly important due to the need to secure access to reserves	Economic liberalism; regional identity unimportant	Not important	Not very important	Similar political problems	Extremely important participation of international financial organisms (IMF and WB)	Important due to extreme asymmetry in oil development capacities (full dependence in STP)	Unimportant

Table 5.1: Explanations for institutional design

Elements of institutional design

The study disaggregated institutional design into four elements or features. I provide below a table with the main variations in these elements uncovered by the empirical chapters.

	Scope	Rules	Norms	Mandate
Middle East	Unintrusive	Weak elaboration; consensus; respect for sovereignty and independence	Strong respect for sovereignty	Distributive
Gulf of Mexico	Unintrusive	Substantial elaboration; consensus over a small part of the territory; respect for sovereignty and jurisdictional issues	Strong influence of UN principles for international cooperation	Distributive
Africa	Intrusive; imposition of legal implementations if necessary	Thoroughly elaborated; high legality and delegation	Strong influence of economic development	Partially distributive and partially process-oriented.

Table 5.2: Institutional design

Some propositions

The main conclusions of the first section (i.e. the nature and sources of regime design), can be summed up in the following propositions:

- The more necessary the regimes are for hydrocarbon and economic development, the more intrusive and legalised the regimes will be.
- States with long traditions in oil development are more respectful of territorial sovereignty.
- Post-colonial governments' cooperation is not based on identity, but on economic ideology and interest.

Nature of cooperation

In the introduction, several possible indicators of the nature of cooperation were provided, including: 1) the degree of institutionalisation and legalisation; 2) the degree of normative and preference change; 3) the degree of policy convergence among actors; 4) the different routes to the above changes; 5) the degree of adjustment of previous policies; 6) the degree to which the institution (or agents active in the institutions) achieved set goals. There are three major conclusions which can be put forward:

First, the design of the regime does affect the nature of cooperation, especially when it comes to the degree of realisation of their initial goals.

Secondly, it can be asserted that in the three cases where the regime was thoroughly elaborated and there was a true intention to achieve development, internal consequences emerged. In the case of the US–Mexico agreement, the initiatives were based on the desire to implement a treaty for the exploration of further, deeper sea areas. In the case of Nigeria and STP, a highly legalised process was conducted between the two countries under pressure from IOCs.

Finally, agreements concluded in the interests of genuine and ongoing development of the HCPs, established within a climate of open discussion, evolved additional positive outcomes to those foreseen in the agreement. This can be seen in the case of Nigeria and STP.

8 Agenda for further research

When I conceived this work, I saw a set of analyses opening the door to the future exploration. The world is increasingly plagued with cases of similar problems regarding sovereignty over transboundary natural resources. The following are but only some immediately apparent avenues for further research, though others will also exist: 1) compare the efficacy and efficiency of development in offshore cases and land-based cases; 2) compare the features of institutional design and later practice of cooperation in cases of shared hydrocarbon and fisheries development; 3) compare the design and nature of cooperation in cases where armed conflict preceded the agreement; and 4) compare the cooperation efficiency based on particular design features.

As I mentioned, these and other research directions could, in some cases, facilitate our understanding of historical disputes as well as inform the design of future regimes.

Buenos Aires, March 2016

BIBLIOGRAPHY

Books

ANTHONY GIDDENS, THE NATIONAL STATE AND VIOLENCE (1985)

ARND BERNAERTS, BERNAESTS' GUIDE TO THE 1982 UNITED NATIONS CONVENTION ON THE LAW OF THE SEA, 36, 42 (1988)

BADR EL DIN A. IBRAHIN, ECONOMIC COOPERATION IN THE ARAB GULF: ISSUES IN THE ECONOMIES OF THE ARAB GULF COOPERATION COUNCIL STATES (2007)

BOUNDARIES AND ENERGY: PROBLEMS AND PROSPECTS (Gerald Blake et al. eds., 1998)

CHARLES TILLY, THE FORMATION OF NATIONAL STATES IN EUROPE (1975)

CRAFTING COOPERATION (Acharya and Johnston eds., 2007)

DODGE, TOBY AND HIGGOT, RICHARD, GLOBALIZATION AND THE MIDDLE EAST: ISLAM, ECONOMY, SOCIETY AND POLITICS (2002).

EDGAR W. BUTLER, JAMES B. PICK and W. JAMES HETTRICK, MEXICO AND MEXICO CITY IN THE WORLD ECONOMY (2001)

EYAL BENVENISTI, SHARING TRANSBOUNDARY RESOURCES. INTERNATIONAL LAW AND OPTIMAL RESOURCE USE (2002)

FRANCISCO PARRA, OIL POLITICS: A MODERN HISTORY OF PETROLEUM (2004).

FRED HALLIDAY, THE MIDDLE EAST IN INTERNATIONAL RELATIONS: POWER, POLITICS AND IDEOLOGY (2005)

HUMAN RIGHTS IN NATURAL RESOURCE DEVELOPMENT (Zillman et al. eds.) (2005)

JONATHAN GRAUBART, LEGALIZING TRANSNATIONAL ACTIVISM: THE STRUGGLE TO GAIN SOCIAL CHANGE FROM NAFTA'S CITIZEN PETITIONS (2008)

JOSEPH A. PRATT, TYLER PRIEST and CHRISTOPHER J. CASTANEDA, OFFSHORE PIONEERS: BROWN & ROOT AND THE HISTORY OF OFFSHORE OIL AND GAS (1997)

KING, KEOHANE AND VERBA, DESIGNING SOCIAL ENQUIRY. SCIENTIFIC INFERENCE IN QUALITATIVE RESEARCH (1994)

MAX SIOLLUN. OIL, POLITICS AND VIOLENCE (2009)

MICHAEL S. CASEY, THE HISTORY OF KUWAIT (2007)

M. M. OLISA, NIGERIAN PETROLEUM LAW AND PRACTICE (1997)

177

NICHO SCHRIJVER, SOVEREIGNTY OVER NATURAL RESOURCES 120 (1997)

NAFTA AS A MODEL OF DEVELOPMENT: THE BENEFITS AND COSTS OF MERGING HIGH AND LOW WAGE AREAS 103 (Richard S. Belous and Jonathan Lemco eds., 1995)

OCEAN DEVELOPMENT AND INTERNATIONAL LAW (2005)

PRINCETON LYMAN. NIGERIA´S ECONOMIC PROSPECTS: AN INTERNATIONAL PERSPECTIVE (2004)

RACHEL BRONSON, THICKER THAN OIL: AMERICA´S UNEASY PARTNERSHIP WITH SAUDI ARABIA (2006)

REGIONAL ATLAS ON WEST AFRICA (Lurent Bossard ed., 2009)

ROBERT T. MORAN, UNITING NORTH AMERICA BUSINESS – NAFTA: BEST PRACTICES (2002)

SOALA ARIWERIOKUMA. THE POLITICAL ECONOMY OF OIL AND GAS IN AFRICA: THE CASE OF NIGERIA (2009)

TAHIR HUSAIN, KUWAITI OIL FIRES: REGIONAL ENVIRONMENTAL PERSPECTIVES (1995)

THE NEW ENERGY PARADIGM (Helm et al. eds., 2007)

THE PEACEFUL MANAGEMENT OF TRANSBOUNDARY RESOURCES (Gerald H. Blake et al. eds., 1995)

THE RATIONAL DESIGN OF INTERNATIONAL INSTITUTIONS (Koremenos et al. eds., 2004)

VALÉRIE MARCEL, OIL TITANS (2006)

WAYNE H. BOWEN, THE HISTORY OF SAUDI ARABIA (2008)

Works in collections

Anne-Marie Slaughter, *International law and international relations theory: a prospectus*, *in* THE IMPACT OF INTERNATIONAL LAW ON INTERNATIONAL COOPERATION at 27, n.30 (Eyal Benvenisti and Moshe Hirsch eds., Cambridge University Press 2004)

James Fearon and Alexander Wendt, *Rationalism v. Constructivism: A Skeptical View*, *in* HANDBOOK OF INTERNATIONAL RELATIONS (2007) 52, 54.

Barry Barton, Catherine Redgwell, Anita Ronne, and Ronald N. Zillman, *Introduction*, *in* ENERGY SECURITY. MANAGING RISK IN A DYNAMIC LEGAL AND REGULATORY ENVIRONMENT 3, 5 (Barry Barton et al. eds., 2004)

Beth A. Simmons and Lisa L. Martin, *International Organizations and Institutions, in* INTERNATIONAL LAW AND INTERNATIONAL RELATIONS, 195.

Stephen D. Krasner, *Structural Causes and Regime Consequences: Regimes as Intervening Variables, in* INTERNATIONAL LAW AND INTERNATIONAL RELATIONS, 3.

James Fearon and Alexander Wendt, *Rationalism v. Constructivism: A Skeptical View*, *in* HANDBOOK OF INTERNATIONAL RELATIONS (2007) 52, 54.

Kenneth W. Abbott, Robert O. Keohane, Andrew Moravcsik, Anne-Marie Slaughter, and Duncan Snidal, The *Concept of Legalization, in* INTERNATIONAL LAW AND INTERNATIONAL RELATIONS, 115

Kenneth W. Abbot and Duncan Snidal, *Pathways to International Cooperation, in* THE IMPACT OF INTERNATIONAL LAW AND INTERNATIONAL COOPERATION, 50.

Beth A. Simmons and Lisa L. Martin, *International Organizations and Institutions, in* INTERNATIONAL LAW AND INTERNATIONAL RELATIONS, 195.

James Fearon and Alexander Wendt, *Rationalism v. Constructivism: A Skeptical View, in* HANDBOOK OF INTERNATIONAL RELATIONS (2007) 52, 54.

Mouton, *The Continental Shelf,* 85 RECUEIL DES COURS (1954, 1) 347, 422.

Rainer Lagoni, *Interim Measures pending Maritime Delimitation Agreements,* 78 Am. J. Int'l Law (1984) 345, 357.

Charles Robson, *Transboundary Petroleum Reservoirs: Legal Issues and Solutions, in* THE PEACEFUL MANAGEMENT OF TRANSBOUNDARY RESOURCES 3, 3 (Gerald H. Blake et al. Eds., 1995)

Gerald H. Blake and R.E. Swarbrick, *Hydrocarbons and International Boundaries: A Global Overview, in* BOUNDARIES AND ENERGY: PROBLEMS AND PROSPECTS 3, 14 (Gerald Blake et al. eds., 1998)

Ibrahim F. I. Shihata and William T. Onorato, *Joint Development of International Petroleum Resources in Undefined and Disputed Areas, in* BOUNDARIES AND ENERGY: PROBLEMS AND PROSPECTS 433, 433(Gerald Blake et al. eds., 1998)

Masahiro Miyoshi, *International Maritime Boundaries and Joint Development: A quest for a Multilateral Approach, in* BOUNDARIES AND ENERGY: PROBLEMS AND PROSPECTS 453, 453 (Gerald Blake et al. eds., 1998).

D. H. Anderson, *Strategies for Dispute Resolution: Negotiating Joint Agreements, in* BOUNDARIES AND ENERGY: PROBLEMS AND PROSPECTS 473 (Gerald Blake et al. eds., 1998).

Rodman R. Bundy, *Natural Resource Development (Oil and Gas) and Boundary Disputes, in* THE PEACEFUL MANAGEMENT OF TRANSBOUNDARY RESOURCES 23 (Gerald H. Blake et al. eds., 1995).

Ian Townsend-Gault and William Stormont, *Offshore Petroleum Joint Development Arrangements: Functional Instrument? Compromise? Obligation?, in* THE PEACEFUL MANAGEMENT OF TRANSBOUNDARY RESOURCES 51, 58 (Gerald H. Blake et al. eds., 1995)

Peter R. Odell, *Hydrocarbons: The Pace Quickens, in* 28, 37 BOUNDARIES AND ENERGY: PROBLEMS AND PROSPECTS 453, 453 (Gerald Blake et al. eds., 1998).

Keith Highet, *New Courts and Old, Old Law and New, and Problems to Come, in* BOUNDARIES AND ENERGY: PROBLEMS AND PROSPECTS 415 (Gerald Blake et al. eds., 1998).

Richard E. Swarbrik, *Oil and Gas Reservoirs Across Ownership Boundaries: The Technical Basis for Apportioning Reserves, in* THE PEACEFUL MANAGEMENT OF TRANSBOUNDARY RESOURCES 41, 47 (Gerald H. Blake et al. eds., 1995).

Clive Schofield, *Blurring the Lines? Maritime Joint Development and the Cooperative Management of Ocean Resources, in* FRONTIER ISSUES IN OCEAN LAW: MARINE RESOURCES, MARITIME BOUNDARIES, AND THE LAW OF THE SEA (The Berkeley Electronic Press, 2009)

Luis E. Rodriguez-Rivera, *Joint Development Zones and other cooperative Management Efforts Related to Transboundary Maritime Resources: A Caribbean and Latin American Model for Peaceful Resolution of Maritime Boundary Disputes, in* FRONTIER ISSUES IN OCEAN LAW: MARINE RESOURCES, MARITIME BOUNDARIES, AND THE LAW OF THE SEA (The Berkeley Electronic Press, 2009).

Stephen D. Krasner, *Structural Causes and Regime Consequences: Regimes as Intervening Variables, in* INTERNATIONAL LAW AND INTERNATIONAL RELATIONS 3 (Beth A. Simmons and Richard H. Steinberg eds., 2006).

Paul F. Diehl, Charlotte Ku and Daniel Zamora, *The Dynamics of International Law: The Interaction of Normative and Operating Systems, in* INTERNATIONAL LAW AND INTERNATIONAL RELATIONS, 426 (Beth A. Simmons and Richard H. Steinberg eds., 2006)

G. H. Blake and R. E. Swarbrick, *Hydrocarbons and International Boundaries: A Global Overview, in* BOUNDARIES AND ENERGY: PROBLEMS AND PROSPECTS 3, 14 (Gerald Blake et al. eds., 1998)

Mohamed Abdullah Al Roken, *Dimensions of the UAE–Iran Dispute over Three Islands*, in UNITED ARAB EMIRATES: A NEW PERSPECTIVE (IBRAHIM AL ABED and PETER HELLYER eds., 2007)

Bassam Fattouh, *OPEC pricing Power: The Need for a New Perspective*, in THE NEW ENERGY PARADIGM (Dieter Helm ed., 2007) 280.

Isaac Cohen, *The NAFTA's Winners and Losers: A Focus on Investment, in* NAFTA AS A MODEL OF DEVELOPMENT: THE BENEFITS AND COSTS OF MERGING HIGH AND LOW WAGE AREAS 37 (Richard S. Belous and Jonathan Lemco eds., 1995)

Harley Shaiken, *The NAFTA, a Social Charter, and Economic Growth*, in NAFTA AS A MODEL OF DEVELOPMENT: THE BENEFITS AND COSTS OF MERGING HIGH AND LOW WAGE AREAS 27 (Richard S. Belous and Jonathan Lemco eds., 1995)

Jonathan I. Charney, *The American Society of International Law Maritime Boundary Project*, 5 MARITIME BOUNDARIES 10 (Gerald H. Blake ed., 1994).

Bosworth, Barry, P. and Susan M. Collins. *The Empirics of Growth: An Update, in* BROOKINGS PAPERS ON ECONOMY ACTIVITY (1996) 113-179

Periodical Materials

David M. Ong, Joint Development of Common Offshore Oil and Gas Deposits: "Mere" State Practice or Customary International Law?, 93 Am. J. INT'L L.771, n. 7.

Ely, The Conservation of Oil, 51 Harv. L. Rev. 1209 (1937–1938)

William T. Onorato, Apportionment of an International Common Petroleum Deposit, 17 INT'L & COMP. L.Q. 85 (1968)

William T. Onorato, Apportionment of an International Common Petroleum Deposit, 26 INT'L & COMP. L.Q. 324 (1977)

J. E. Horigan, Unitization of Petroleum Reservoirs Extending Across Sub-sea Boundary Lines of Bordering States in the North Sea, Natural Resources Lawyer, Vol. 7, No. 1, Winter 1974

Jacobs, Unit Operation of Oil and Gas Fields, 57 YALE L.J. 1207 (1947–1948)

Talaat El Ghoneimy, The Legal Status of the Saudi-Kuwait Neutral Zone, 15 INT'L & COMP. L.Q. 690 (1966)

Hosni, The Partition of the Neutral Zone, 60 AJIL 735 (1966)

Al-Baharna, A Note on the Kuwait-Saudi Arabia Neutral Zone Agreement of July 7, 1965, Relating to the Partition of the Zone, 17 INT'L & COMP. L.Q. 730 (1968)

Utton, Institutional Arrangements for Developing North Sea Oil and Gas, 9 Virginia J. INT'L L. 66 (1968–1969)

Beth A. Simmons and Lisa L. Martin, International Organizations and Institutions, in INTERNATIONAL LAW AND INTERNATIONAL RELATIONS, 195

Mouton, The Continental Shelf, 85 RECUEIL DES COURS (1954, 1) 347, 422

Rainer Lagoni, Interim Measures pending Maritime Delimitation Agreements, 78 Am. J. Int'l Law (1984) 345, 357

Churchill, R.R., 'Falkland Islands – Maritime Jurisdiction and Cooperative Arrangements with Argentina,' Current Legal Developments, 46 International and Comparative Law Quarterly 463–477 (1997)

Alberto Szekely, The International Law of Submarine Transboundary Hydrocarbon Resources: Legal Limits to Behavior and Experiences for the Gulf of Mexico, 26 Natural Resources Journal 766 (1986)

David M. Ong, Joint Development of Common Offshore Oil and Gas Deposits: "Mere" State Practice or Customary International Law?, 93 Am. J. INT'L L.771, 798

Schofield, C.H., Unlocking the Seabed Resources of the Gulf of Thailand, 29 Contemporary

Southeast Asia 286-308 (August 2007)

Richard J. McLaughlin, Hydrocarbon Development in the Ultra-Deepwater Boundary Region of the Gulf of Mexico: Time to Reexamine a Comprehensive US-Mexico Cooperation Agreement, 39 Ocean Development and International Law 1 - 31 (2008)

Finnemore, Martha and Sikkink, Kathryn, International Norm Dynamics and Political Change, 52 International Organizations 887–917 (1998)

Alan K. Henrikson, Facing across Borders: The Diplomacy of Bon Voisinage, 21 International Political Science Review 121–147 (Apr., 2000)

Zuhayr Mikdashi, Cooperation among Oil Exporting Countries with Special Reference to Arab Countries: A Political Economy Analysis, 28 and 32 International Organization, 1

Euclid A. Rose, OPEC's Dominance of the Global Oil Market: The Rise of the World's Dependency on Oil, 58 MIDDLE EAST JOURNAL 424

Gerd Nonneman, Saudi-European Relations 1902-2001: A Pragmatic Quest for Relative Autonomy, 77 INTERNATIONAL AFFAIRS 631–661

Ali R. Abootalebi, Middle East Economies: a Survey of Current Problems and Issues, 3 Middle East Review of International Affairs (1999)

Alexandre Oliveira, Innovation urged in NOC, IOC relations with supplier, OIL AND GAS JOURNAL (Week of August 3, 2009)

Key World Energy Statistics 2008, INTERNATIONAL ENERGY AGENCY (2009) 11

Gwenn Okruhlik and Patrick J. Conge, The Politics of Border Disputes: On the Arabian Peninsula, 54 Int'l Jour. 230

George W. Downs, David M. Rocke, and Peter N. Barsoom, Managing the Evolution of Multilateralism, 52 INT'L ORG 397–419 (1998)

Christian Reus-Smit, The Constitutional Structure of International Society and the Nature of Fundamental Institutions, 51 INT'L ORG 569 (1997)

Kenneth W. Abbott and Duncan Snidal, Why States Act Through Formal International Organizations, 42 JOURNAL OF CONFLICT RESOLUTION 24 (1998)

Jorge A. Vargas, Mexico's Legal Regime over its Marine Spaces: A proposal for the Delimitation of the Continental Shelf in the Deepest Part of the Gulf of Mexico, 26 U. MIAMI INTER-AM. L. REV, 189, 238 (1995)

Mark B. Feldman and David Colson, The Maritime Boundaries of the United States, 75, AM. J. INT'L. 729, 730 (1981)

Jorge A. Vargas, US Marine Scientific Research Activities Offshore Mexico: An Evaluation of Mexico's Recent Regulatory Legal Framework, 24 DENV. J. INT'L L. & POL'Y I, 36 (1995)

David B. Sheinbein, Delimitation of Western Gap Land in the Gulf of Mexico: A Need for Diplomatic Resolution, 6 TUL. J. INT'L & COMP. L. 583, 595.

US – Mexico gulf treaty pressures rising, OIL & GAS J., May 12, 1997

Joe C. Ashby, Labor and the Theory of the Mexican Revolution under Lázaro Cárdenas, 20 THE AMERICAS (1963), 158–199

Gerald Karey, US Mexican Officials Agree Doughnut Hole Boundary, PLATT'S OILGRAM NEWS, June 12, 2000

US, Mexico Soon to Delineate Gulf 'Gap', OIL & GAS JOURNAL, Dec. 15, 1997, at 27–28.

Ronald Buchanan, Gulf 'Donut Hole' Focus of Study by PEMEX, PLATT'S OIL-GRAM NEWS, May 22, 1998

Mathew Robinson and Eric Kronenwetter, Dumping Motion Fifed: Mexico Eyes Ratification, OIL DAILY, June 30, 1999.

US Mexico Complete Gulf of Mexico Treaty Ratification, PLATT'S OILGRAM NEWS, January 19, 2001

Deberán informar de hoyos de dona, EXELCIOR, 10 de julio de 2008. (They will have to inform on doughnut holes, EXELCIOR JOURNAL, July 10, 2008)

Michael Klare and Daniel Volman, The African 'Oil Rush' and US National Security, 27 Third World Quarterly (2006) 609–628

Michael Klare and Daniel Volman, America, China & the Scramble for Africa's Oil, 33 Review of African Political Economy (2006), 297–309

NIGERIA – The Petroleum Reserves, APS REVIEW GAS MARKET TRENDS, Monday, July 30, 2007, available at http://www.allbusiness.com/agriculture-forestry-fishing-hunting/support-activities/4493935-1.html

Nigeria – History of Oil Exploration, APS REVIEW GAS MARKET TRENDS, Monday, August 3 2009, available at http://www.allbusiness.com/mining-extraction/oil-gas-exploration-extraction-oil-oil/12634580-1.html

NIGERIA – The OPEC Decision Makers – Part 9, APS DIPLOMAT OPERATIONS OIL DIPLOMACY, Monday, September 18, 2000, available at http://www.allbusiness.com/government/714853-1.html

Nigeria Decision Makers – The NNPC Structure, APS Review Downstream Trends, Monday, August 20, 2007, available at http://www.allbusiness.com/trade-development/international-trade-exports-imports-by/5495884-1.html

FT Africa Oil and Gas, Financial Times, March 1, 2006.

Bosworth, Barry, P. and Susan M. Collins. The Empirics of Growth: An Update, in BROOKINGS PAPERS ON ECONOMY ACTIVITY (1996) 113–179

Neil Ford, Where does all the oil money go?, AFRICAN BUSSINESS, Thursday, June 1, 2006. Available at http://www.allbusiness.com/accounting/accounts-receivable/4097569-1.html

Beth Anne Wilson and Geoffrey N. Keim, India and the Global Economy, 41 Business Economics (2006) 28–36

NIGERIA – Foreign Partners' Role In Deep-Water Oil, APS REVIEW OIL MARKET TRENDS, Monday, August 8, 2005. Available at http://www.allbusiness.com/mining/oil-gas-extraction-crude-petroleum-natural/494833-1.html

IMF Executive Board Approves a Two-Year Policy Support Instrument for Nigeria, IMF Press Release, available at www.imf.org/external/pubs/ft/scr/2005/cr05229.htm

Nigeria – Part 2 – The Oil & Gas Fields And Foreign Operators, APS REVIEW OIL MARKET TRENDS, Monday, August 10, 2009, available at http://www.allbusiness.com/africa/1011432-1.html;

NIGERIA – Foreign Partners' Role In Deep-Water Oil, APS REVIEW OIL MARKET TRENDS, Monday, August 8, 2005, available at http://www.allbusiness.com/minin g/oil-gas-extraction-crude-petroleum-natural/494833-1.html

Temitope W. Oshikoya, Nigeria in the global economy: Nigeria's integration into the global economy is below its potential, BUSSINESS ECONOMICS, Tuesday, January 1, 2008, available at http://www.allbusiness.com/economy-economic-indicator s/output-demand-gross/11408248-1.html

Temitope W. Oshikoya, Nigeria in the global economy: Nigeria's integration into the global economy is below its potential, BUSSINESS ECONOMICS, Tuesday, January 1, 2008, available at http://www.allbusiness.com/economy-economic-indicator s/output-demand-gross/11408248-1.html

Neil Ford, Where does all the oil money go?, AFRICAN BUSSINESS, Thursday, June 1, 2006, available at http://www.allbusiness.com/accounting/accounts-receivable/40 97569-1.html

Nigeria – Nigerian Politicians & IOC Attack New Petroleum Reforms Bill, APS REVIEW OIL MARKET TRENDS, Monday, August 3, 2009, available at http://w ww.allbusiness.com/government/government-bodies-offices-legislative/12634621-1 .html

QATAR – Oil & Gas Fields – The Offshore Fields, APS Review Gas Market Trends (2003), http://www.allbusiness.com/mining/oil-gas-extraction-crude-petroleum-nat ural/635460-1.html;

KUWAIT – The Offshore Fields, APS Review Gas Market Trends (2005), http://www. allbusiness.com/mining/oil-gas-extraction-crude-petroleum-natural/437634-1.html;

Edward Burton, Saudi Arabia aims to bolster offshore oil and gas program, Offshore (2009), http://www.allbusiness.com/mining-extraction/oil-gas-exploration-extractio n-oil-oil/13167527-1.html;

Qatar's Offshore Oil & Gas Fields, APS Review Gas Market Trends (2009), http://ww w.allbusiness.com/mining-extraction/oil-gas-exploration-extraction-oil-oil/1285908 6-1.html

Middle East Economic Digest (January 2002)

Treaties

THE GENEVA CONVENTION ON THE CONTINENTAL SHELF, U.N. Doc. A/ Conf. 13/L.55

AGREEMENT CONCERNING THE DELIMITATION OF THE CONTINENTAL SHELF IN THE PERSIAL GULF, Saudi Arabia – Bahrain, Feb. 22, 1958, NATIONAL LEGISLATION AND TREATIES RELATING TO THE LAW OF THE SEA 409, U.N. Doc. ST/LEG/SER.B/16

AGREEMENT CONVERNING THE WORKING OF COMMON DEPOSITS OF NATURAL GAS AND PETROLEUM, Czech Rep. – Aus., Jan. 23, 1960, 495 UNTS 134

SUPPEMENTARY AGREEMENT CONCERNING ARRANGEMENTS FOR COOP-ERATION IN THE ELMS ESTUARY, Neth. – FRG, May. 14, 1962 509 UNTS 140

AGREEMENT RELATING TO THE PARTITION OF THE NEUTRAL ZONE, Kuwait – Saudi Arabia, Jul. 7, 1965, 4 ILM 1134 (1965)

AGREEMENT ON THE SETTLEMENT OF MARITIME BOUNDARY LINES AND SOVEREIGN RIGHTS OVER ISLANDS, Qatar – Abu Dhabi, Mar. 20, 1969, LEGISLATION AND TREATIES 403

MEMORANDUM OF UNDERSTANDING, Iran–Sharjah, Nov. 18, 1971, reprinted in ALI A. EL-HAKIM, THE MIDDLE EASTERN STATES AND THE LAW OF THE SEA 208 (1979)

CONVENTION ON THE DELIMITATION OF THE CONTINENTAL SHELF IN THE BAY OF BISCAY, Fr. Spain, Jan. 29, 1974, LEGISLATION AND TREATIES 445

AGREEMENT CONCERNING JOINT DEVELOPMENT OF THE SOUTHERN PART OF THE CONTINENTAL SHELF ADJACENT TO THE TWO COUN- TRIES, Japan – South Korea, Feb. 5, 1974, in 4 NEW DIRECTIONS IN THE LAW OF THE SEA 117 (R. R. Churchill & Myron H. Nordquist eds., 1975)

AGREEMENT RELATING TO THE EXPLOITATION OF THE SEA-BED AND SUBSOIL OF THE RED SEA IN THE COMMON ZONE, Sudan – Saudi Arabia, May 16, 1974, LEGISLATION AND TREATIES, at 452, U.N. Doc. ST/LEG/ SER.B/18, U.N. Sales No. E/F.76.V.2 (1976)

AGREEMENT RELATING TO THE EXPLOITATION OF THE FRIGG RESER- VOIR AND THE TRANSMISSION OF GAS THEREFROM TO THE UNITED KINGDOM, United Kingdom – Norway, May 10, 1976, 1098 UNTS.

UNITED NATIONS CONVENTION FOR THE LAW OF THE SEAS, 1833 UNTS 397, reprinted in UNITED NATIONS, OFFICIAL TEXT OF THE UNITED NATIONS CONVENTION ON THE LAW OF THE SEA WITH ANEXES AND INDEX, UN Sales No. E.83.V.5 (1983) (entered into force Nov. 16, 1994)

Agreement between the Government of the United Kingdom of Great Britain and Northern Ireland and the Government of the Kingdom of Norway Relating to the Delimitation of the Continental Shelf Between the Two Countries (London 10 March 1965, 551 UNTS 214).

Treaty between Saudi Arabia and Bahrain (1958); Treaty between Saudi Arabia and Qatar (1965); Treaty between Saudi Arabia and Iran (1968); Treaty between Iran and Bahrain (1971),

Available at http://www.un.org/Depts/los/LEGISLATIONANDTREATIES/persian_gu lf.htm

Exchange of notes between Saudi Arabia and the UAE concerning the Joint Minutes on the land and maritime boundaries to the Agreement of 4 December 1965 between the State of Qatar and the Kingdom of Saudi Arabia on the delimitation of the offshore and land boundaries, available at http://www.un.org/Depts/los/LEGISL ATIONANDTREATIES/STATEFILES/SAU.htm.

Treaty to Resolve Pending Boundary Differences and Maintain the Rio Grande and Colorado River as the International Boundary, 23 November 1970 (entry into force: 18 April 1972; registration #: 11873; registration date: 17 July 1972)

BIBLIOGRAPHY

Maritime Boundaries Agreement Effected by Exchange of Notes between the United States of America and Mexico 24 November 1976

Treaty on maritime boundaries between the United States of America and the United Mexican States (Caribbean Sea and Pacific Ocean), 4 May 1978 (entry into force: 13 November 1997; registration #: 37399; registration date: 12 April 2001)

Treaty between the Government of the United States of America and the Government of the United Mexican States on the Delimitation of the Continental Shelf in the Western Gulf of Mexico beyond 200 Nautical Miles, 9 June 2000 (entry into force: 17 January 2001; registration #: 37400; registration date: 12 April 2001).

Treaty on Marine Boundaries Between the United States of America and the United Mexican States, May 4, 2978, US – Mex., 17 L.L.M 1073.

Treaty Between the Government of the United States of America and the Government of the United Mexican States on the Delimitation of the Continental Shelf in the Western Gulf of Mexico Beyond 200 Nautical Miles, June 9, 2000, US – Mex., 8 TREATY DOC. No. 106-39 (2000).

Agreement between the Government of the United Kingdom of Great Britain and Northern Ireland and the Government of the Kingdom of Norway Relating to the Delimitation of the Continental Shelf Between the Two Countries (London 10 March 1965, 551 UNTS 214).

Exchange of notes between Saudi Arabia and the UAE concerning the Joint Minutes on the land and maritime boundaries to the Agreement of 4 December 1965 between the State of Qatar and the Kingdom of Saudi Arabia on the delimitation of the offshore and land boundaries. Available at http://www.un.org/Depts/los/LEGIS LATIONANDTREATIES/STATEFILES/SAU.htm.

Legislative materials and reports

ILC Report A/65/10, 2010, Chp. XII, paras. 374-384

GA Res. 2995 (XXVII)

GA Res. 2996 (XXVII)

GA Res. 3129 (XXXVIII)

UNEP Offshore Oil and Gas Environment Forum, http://www.natural-resources.org/of fshore/background/bgnote.htm

WORLD TRADE ORGANIZATION (2005), *available at* www.stat.wto.org/CountryP rofiles

Regional Economic Outlook. Sub-Saharan Africa, IMF (2006), also, *World Economic Outlook*, IMF (2006)

World Economic Outlook, IMF (2006)

Key World Energy Statistics *2008*, INTERNATIONAL ENERGY AGENCY (2009) 11

2006 Budget Speech, FEDERAL GOVERNMENT OF NIGERIA (2006), *available at* www.budgetoffice.gov.ng

US Report of the National Energy Policy Development Group (2001)

National Strategy for Maritime Security, DEPARTMENT OF DEFENSE AND HOMELAND SECURITY (2005)

ENERGY INFORMATION ADMINISTRATION, OFFICE OF ENERGY MARKETS AND END USE, INTERNATIONAL STATISTICS DATABASE AND INTERNATIONAL ENERGY ANNUAL 1997, DOE/ EIA-0219(97) (April, 1999)

Secretaría de Energía del Ministerio de Economía, Comité de Información, RES: CI-IV-2008-019, at 1.

Doing Business in 2006: Creating Jobs, THE WORLD BANK AND INTERNATIONAL FINANCE (2006), available at http://www.doingbusiness.org/Downloads/

Nigeria Private Sector Assessment, THE WORLD BANK REGIONAL PROGRAM ON ENTERPRISE DEVELOPMENT, *available at* http://www.usaid.gov/ng/downloads/reforms/nigeriaprivatesectorassessmenttechnicalpapers.pdf

Nigeria Private Sector Assessment, THE WORLD BANK REGIONAL PROGRAM ON ENTERPRISE DEVELOPMENT, *available at* http://www.usaid.gov/ng/downloads/reforms/nigeriaprivatesectorassessmenttechnicalpapers.pdf

A. Kraay Kaufmann and M. Mastruzzi, *Governance Matters IV: Governance Indicators for 1996-2004*, THE WORLD BANK (2005), *available at* www.worldbank.org/wbi/governance/pubs/govnatters4.html

Tumsah, K.M., *Nigeria – DRSTP Joint Development Zone: A Legal Perspective*, 1 JDZ News (2005) 8.

2006 Budget Speech, FEDERAL GOVERNMENT OF NIGERIA (2006), *available at* www.budgetoffice.gov.ng

Cases

CASE CONCERNING MARITIME IDELIMITATION AND TERRITORIAL QUESTIONS (Qatar v. Bahrain), 1994 I.C.J. (Jurisdiction and Admissibility) (July 1)

NORTH SEA CONTINENTAL SHELF (Federal Republic of Germany v. Denmark), 1969 I.C.J. (Judgment) (February 20)

NORTH SEA CONTINENTAL SHELF (Federal Republic of Germany v. Denmark), 1969 I.C.J. (Judgment) (February 20)

187